U.S. INVESTMENT IN THE FOREST-BASED SECTOR IN LATIN AMERICA

U.S. INVESTMENT IN THE FOREST-BASED SECTOR IN LATIN AMERICA

Problems and Potentials

HANS M. GREGERSEN and ARNOLDO CONTRERAS

Published for Resources for the Future, Inc.
By The Johns Hopkins University Press
Baltimore and London

This report is part of the RFF Latin American Program, which is directed by Pierre R. Crosson. The figures for this book were drawn by Frank and Clare Ford. The book was edited by Jo Hinkel.

RFF editors: Mark Reinsberg, Joan R. Tron, Ruth B. Haas, Jo Hinkel

Copyright © 1975 by The Johns Hopkins University Press
All rights reserved
Manufactured in the United States of America

Library of Congress Catalog Card Number   74-21754

ISBN 0-8018-1704-8

Library of Congress Cataloging in Publication Data will be found on the last printed page of this book.

Acknowledgments

A great number of individuals and organizations gave of their time and knowledge to make this study possible, and we take this opportunity to sincerely thank them—all those officials of Latin American countries, corporate officers here and abroad, U.S. government and international agency personnel, and others who so generously gave us their cooperation and put up with our sometimes overly detailed and seemingly obvious questions. The study could not have been carried out without their help.

Financial support for the study was generously given by Resources for the Future, Inc., and we are deeply thankful to them. Naturally, their financial support does not necessarily imply support of the ideas expressed in the study. We also thank the University of Minnesota for its cooperation in giving us the necessary time and aid to complete the study. And we particularly thank Marilyn Workman for her valiant efforts in typing and retyping the manuscript.

Pierre Crosson of RFF and Michael Nelson, program advisor in agriculture for the Ford Foundation in Mexico, provided detailed review and criticism of the manuscript. In addition, John Zivnuska, Bill Hyde, Gordon Fox, Mike Arnold, Peter Arnold, and a great many other individuals provided valuable suggestions for additions and changes. We thank them all for their time and excellent comments, but hasten to dissociate them from any of the shortcomings of the study.

St. Paul, Minnesota

H. M. G.
A. C.

Contents

Acknowledgments   v

One **Introduction**   1
    Foreign Investment in the Forest-Based
       Sector of Less-Developed Countries   4
    Approach of This Study   6

Two **Forests and Forest-Based Activity in**
    Latin America   11
    Forest Resources of Latin America   11
    Forest Industries and Production   14
    Trade in Forest Products   15
    Market Size   15

Three **Host Country and Foreign Investor**
    Motivations   23
    Host Country Objectives and Motivations   23
    Foreign Investor Objectives and Motivations 29

Four **U.S. Direct Private Investment**   33
    Contribution to Income and Employment   35
    Foreign Exchange Contribution   38
    Contribution to Development of Technical
       Forestry   40
          Natural Coniferous Forests   40
          Forest Plantations   41
          Natural Tropical Hardwood Forests 42

Five **Interaction Between Host Country and**
    Foreign Investor   47
    Interactions in the Foreign Investment
       Process   47

Major Types of Interaction Situations   48
Summary of the Interaction Process   50
The Interaction Process in Latin America 55
Forest Concession and Management
   Contracts   58
   Conditions of Contracts in Latin
      American Countries   59
   Comparisons with Other Regions   62
   Negotiating Agreements—A Summary   63
Why New Investments Are Scarce   64
Failures of Past Projects   70
New Investment in Expansion of
   Existing Operations   72

Six **Implications for the Future**   75
Opportunities for Forest-Based
   Development   75
Overcoming Obstacles to Expanded
   Foreign Involvement   78
Reducing Uncertainties   82
   Uncertainties Related to Host Country
      and Foreign Investor Policies   82
   Reducing Uncertainties Through
      Improved Flow of Information   89
Requirements for Productive Negotiation
   and Administration of Contracts   97
General Implications for Policy   98

References   101

Appendix I. U.S. Direct Private Foreign
   Involvement in the Forest-Based Sector
   of Latin America   107

Appendix II. Interview Schedules   109

Tables

Table 1. Characteristics of Main Projects Studied 9

Table 2. World Production of Forest Products, 1959, 1966, and 1970 16

Table 3. Exports of Major Forest Products, 1966–1970 17

Table 4. Imports of Major Forest Products, 1966–1970 18

Table 5. Total Net Imports and Net Exports of Major Forest Products in Selected Latin American Countries, 1959, 1966, and 1970 19

Table 6. Apparent Consumption of Forest Products in All Major Latin American Countries, 1959, 1966, and 1970 19

Table 7. Apparent Consumption of Forest Products in the United States and All Less-Developed Countries, 1970 20

Table 8. Apparent Consumption of Forest Products in Four Major Latin American Countries, 1970 20

Table 9. Reasons for Host Country Acceptance of Foreign Investment in the Forest-Based Sector 27

Table 10. Local Expenditures of U.S. Majority-Owned, Forest-Based Affiliates in Latin America, 1966 35

Table 11. Number of Local and U.S. Employees in U.S. Majority-Owned, Forest-Based Affiliates in Latin America, 1966 36

Table 12. Labor Requirements for Felling and Cross-cutting of Logs in Natural Tropical Forests 37

Table 13. Derivation of Net Foreign Exchange Effect of U.S. Forest-Based Affiliates in Latin America, 1966 39

Table 14. Duration and Possibility of Renewal of Long-term Contracts as Determined in the Legislation of Various Countries 60

Table 15.   Main   Causes   and   Incidence   of   Failure
            Among Projects Studied   70
Table 16.   Forest-Based Projects Invested in by ADELA
            Investment Company, as of September 30,
            1972   84
Table 17.   Host Government and United Nations Ex-
            penditures on Investment Information in
            Latin America, as of June 1966   91

Figures

1. Basic determinants of foreign investor and host country interaction outcome   49
2. Summary model describing the interaction between foreign investor and host country for a given investment proposal   51
3. Phases in the interaction between host country and foreign investor   53
4. Interaction of host country and foreign investor minimum requirements   80

U.S. INVESTMENT IN THE FOREST-BASED SECTOR IN LATIN AMERICA

chapter one **Introduction**

For the past ten years or so, economists have devoted increasing efforts to studying foreign direct private investment in the less-developed world, focusing particularly on the impact of foreign investment on economic development and the growth of the "multinationals." Most researchers have dealt with the macro aspects of foreign direct investment in less-developed countries (LDCs), while some have looked at the process by which firms make foreign investment decisions, and others have analyzed the foreign investment experience of specific sectors.[1]

Within this latter category, two of the most widely studied sectors are mining and petroleum. This is partly because of their importance in the total foreign investment picture, and partly because foreign investments in these sectors have been quite "visible" and have been associated with much political controversy. These sectors also involve exploitation of nonrenewable natural resources, and it is this common characteristic which provided a basis for the recent RFF-sponsored study by Raymond F. Mikesell and colleagues, dealing with foreign investment in the petroleum and mineral industries, primarily in Latin America.[2]

The present study is an attempt to complement Mikesell's work. Its objective is to provide an analysis of the foreign investment process and experience in a *renewable* natural resource sector—in this case, the forest-based sector.[3] Our principal focus is on Latin America, but occasional reference is made to other areas as well.

The basic difference between nonrenewable and renewable natural resources is that the latter can be regenerated or replaced within a period of time meaningful in the context of human use of such re-

[1] For a review of recent research trends, cf. J. H. Dunning, ed., *International Investment.*
[2] R. F. Mikesell, ed., *Foreign Investment in the Petroleum and Mineral Industries: Case Studies of Investor-Host Country Relations.*
[3] We include forestry, logging, sawnwood, plywood and other wood-based panels, pulp and paper, and miscellaneous wood products.

sources. More important, man can with proper management utilize some renewable resources and at the same time increase their quality and quantity for the future, thereby adding to a more permanent development capital.[4]

An argument commonly cited within the host country (HC) against outside involvement in nonrenewable natural resource exploitation is that the foreign investor (FI) depletes national resources (capital) without leaving adequate compensation in the HC. In the case of renewable resources this particular argument should be avoidable, since the HC, theoretically at least, can design and enforce policies that require renewal and improvement of the resource capital as it is utilized. Realistically, while policies for sustained-yield management and regeneration of the forest resource exist in most concession agreements and forestry laws in the LDCs, many of the forest-based projects in the tropics have in fact been "mining" operations. As we shall see, there are valid technical and economic reasons why this is so in many areas.

Yet, there are also many foreign projects that do involve active management programs to insure the renewal and improvement of the forest resource which provides a basis for the investment. For example, Myint concluded that FIs practiced scientific management of forest resources in Burma and Thailand where adequate administration and policies existed.[5] In these countries during the period preceding World War II, FIs contributed to a buildup of a permanent exploitation and management of teak (*Tectona grandis*). The point is that proper management, where it exists, depends on the adoption and enforcement of policies and on the economic terms on which the forest resources are permitted to be exploited.

Quite aside from the question of renewability, there are a number of different factors associated with the economics of exploitation and development of renewable and nonrenewable natural resources which make for other significant differences in the character of foreign investment projects in the two sectors. For example, only 5 percent of the sales of U.S. foreign affiliates in the paper and allied products subsector in Latin America are exported outside the region as compared with some 80 percent of sales in the foreign mining sector. This difference has various implications, for example, in terms of linkage effects and foreign exchange impact.

---

[4] For a discussion of classification of natural resources and their differences, cf. S. V. Ciriacy-Wantrup, *Resource Conservation: Economics and Policies* (3rd ed.).

[5] H. Myint, *Southeast Asia's Economy*, p. 94.

Most investments in mineral and petroleum exploitation are very large because of the nature of economies of scale in production and other considerations associated with such exploitation. In contrast, investments in forest exploitation and processing may be very small and still economically efficient. The largest forest-based projects are pulp and paper mills, with an investment in plant and equipment of some $40–60 million or so. This is smaller than the average mineral or petroleum project and very large compared with the majority of forest-based projects. Because of their large size relative to total activity in their sector, nonrenewable resource projects tend to be much more "visible"; and in most LDCs, visibility is associated with attacks by nationalistic interests. Associated with this point, there are only some five companies that control the bulk of the foreign investment activity in petroleum in LDCs,[6] as compared with thirty to forty major companies that dominate in the forest-based sector. The petroleum companies are well identified in the LDCs, both through brand name sales (which one does not generally find in the forest-based sector) and through the mass media.

Individual projects in the forest-based sector are smaller, and so is the total foreign involvement in the sector in Latin America. However, our concern is not with these relative magnitudes. The justifications for looking at the forest-based sector relate to its position as one of the most important renewable natural resource sectors in Latin America: the fact that 40–50 percent of the land area of Latin America is classified as forest land and potential developments are untapped, that foreign investment plays a prominent role in the sector, that the potential of the sector in terms of import savings, export earnings, and export diversification are significant, and that the sector is an important and necessary link in providing a broad array of critical inputs in the development process, for example, building materials and paper.[7]

A further justification for the study is the need for more sectoral studies of foreign investment to complement the growing volume of research conclusions relating to the macro aspects of the subject. Empirical studies at the sector and industry levels provide the basis for specific policies needed to effectively guide developments in specific types of activities. Such policies are best developed from sectoral studies which first identify and analyze sector problems and conflicts

---

[6] Mikesell, *Foreign Investment*, p. 9.

[7] The importance and potentials of the sector are indicated elsewhere by many authorities, for example, D. Szabo, "United States Statement on Export Development"; and J. Westoby, "The Role of Forest Industries in the Attack on Economic Underdevelopment."

before offering the possible solutions to such problems. This is even more important where significant differences exist between the sectors. This is not to say that there are no general investment policies which apply to all sectors. On the contrary, there are a great number of such policies which are useful in guiding foreign investment toward achievement of national social and economic objectives such as those related to employment, income growth, balance of payments, and regional development. While we by necessity have to treat the general policies and the conflicts and the problems which they are designed to overcome, our main concern is with the sector-specific conditions, problems, and policies.

Foreign Investment in the Forest-Based Sector of Less-Developed Countries

Foreign investment in the forest-based sector of LDCs dates back a long time, and, at present, almost all the major U.S. corporations in this sector have direct investment interests in one or more of the LDCs. Over the years the interest has shifted from one country to another, in much the same way that investment interests in materials and petroleum shift as new reserves are discovered. For example, a country of current interest because of its resources and government attitude is Indonesia.[8] Not only Americans have flocked there to get concessions, but also Japanese, Filipinos, and others. According to recent official statistics, plans for more than $550 million worth of forest-based projects existed in Indonesia as of 1971.[9]

U.S. foreign investment in some parts of the forest-based sector has been growing steadily in recent years. Majority-owned foreign affiliates in the paper and allied products group accounted for more than $1.5 billion in foreign direct investments in 1966, a 17.1 percent increase over the 1965 level and the highest growth rate for any industry subgroup.[10] About 12 percent of the total investment is in LDCs.

---

[8] One company spokesman told us jokingly that he saw more of his fellow industry representatives when he walked through the lobby of the main hotel in Djakarta than he normally sees at a trade meeting in the United States.

[9] Bank Exspor Impor of Indonesia (n.d.).

[10] U.S. Department of Commerce, "Paper Industry Leads in Foreign Growth." It should be noted that the paper and allied products group includes some industries such as paper conversion which does not include primary processing of wood. However, most of these other activities rely on wood as a primary raw material, with the remainder using nonwood fibers and waste paper. The statistics do not permit us to separate these categories.

A number of the existing U.S. subsidiaries and affiliates in the forest-based sector of Latin America have been operating for many years.[11] It is a fairly stable sector, and there are few companies that have left the region due to direct political problems, although there are a fair number of projects that have failed or not materialized due to a failure of the FI and HC to reach agreement on various points of conflict. Investment in the region over the past five years is characterized by very few commitments to countries other than the ones in which investors are already involved. Most of the recent investment is in expansion of existing foreign affiliates. We will have more to say in Chapters Three and Five on this rather significant aspect of current investment activity. There are at least forty-eight U.S. corporations involved in the forest-based sector of Latin America and nearly one hundred projects (including some which are in the hands of individuals or groups of U.S. citizens).[12]

Not all the direct foreign investment in the region is associated with individuals or individual corporations. For example, ADELA, a multinational investment group in Latin America, including substantial U.S. participation, has about one-fifth of its authorized long-term investments of $103 million in some eleven pulp and paper and wood products projects in Latin America. This is significant considering that ADELA is involved in most sectors.

While the major portion of foreign investment in the forest-based sector of Latin America is accounted for by U.S. investors, there is investment activity by European companies and Japan. For example, agreement was recently reached between Brazilian interests and eleven Japanese companies for a pulp project that will initially produce 250,000 tons a year. This joint venture will supposedly be the largest of its kind in the developing world.[13] The project includes a $1 million study of forest utilization in the states of Minas Gerais and Espírito Santo. Other large Japanese trading companies such as Mitsubishi and Nissho-Iwai are actively involved with plans for forest-based projects in Brazil. European firms such as Bruynzeel, N.V. from Holland, Borregaard from Norway, and others are also involved in Latin America.

[11] A more detailed discussion of the magnitude of U.S. involvement in Latin America is given in Chapter Four.

[12] See Appendix I for a listing of the companies.

[13] The Japanese will put $600 million into the estimated total investment of $1 billion. The project includes planting of about 1 million acres with eucalyptus, enough for 750,000 tons of pulp and 3 million tons of chips per year in about six years from start up.

Approach of This Study

While we are interested in the unique aspects of foreign investment in the forest-based sector in relation to its role as a major renewable natural resource-based sector, we are more generally interested in the nature of its impact on development in Latin America and the performance of the sector in terms of HC and FI objectives and interactions. Indeed, it is necessary to analyze these broader aspects in order to understand those differences associated with the renewability of the resource being utilized.

The basic questions explored are the following:

1. What are the present forest conditions in the region and what is the present situation with regard to production, trade, and consumption of forest products? What is the present involvement of FIs in the sector? Answers to these questions define the basic background for a more detailed analysis of selected foreign investment projects and a means for placing these specific projects in a more general framework.

2. What are the HC motivations and objectives which create an interest in obtaining foreign investment in the forest-based sector?

3. What has been the impact of past foreign investment in this sector on the Latin American countries involved?

4. What are the FI objectives and motivations which make projects in the forest-based sector of Latin America appealing to U.S. corporations, and how do they compare foreign investment opportunities with domestic opportunities?

5. What has been the experience of U.S. investors in the forest-based sector of Latin America? What are the characteristics of successful and unsuccessful projects?

6. What is the nature of the interaction process between FI and HC? What are the problems or conflicts which arise in the process of reconciling HC and FI motivations through negotiation for a project? To what extent are conflicts caused by misunderstanding (or inadequate information) rather than by actual conflicting differences in motivations and/or minimum conditions for investment set by the two?

7. What means have been, are, and could be used to ameliorate conditions or overcome the conflicts which arise?

8. Finally, what are some of the main differences between foreign investment processes and experiences in the forest-based sector and the experience in other sectors?

Our first task in developing answers to these questions is to define the extent of the Latin American forest resource and its current development, along with other production and market factors which might attract foreign investment to the region. This we do in Chapter Two, where the overall trends in the Latin American forest-based sector are briefly reviewed.

In Chapter Three we explore the objectives and motivations which influence the activities and interaction between FI and HC in the forest-based sector. In Chapter Four we describe the actual nature and extent of U.S. involvement in the forest-based sector in Latin American countries, looking at the experience to date in the context of the HC and FI objectives discussed in Chapter Three.

Actual experience with foreign investment projects, as well as the perceived benefits and costs involved, determine the motivation for becoming involved with any given project as well as the nature of the interaction between a given FI and HC. The interaction process in the forest-based sector in Latin America is explored in Chapter Five, which also details some of the conditions other than HC and FI objectives which influence the interaction process and its outcome. We are particularly interested in the reasons behind the lack of new investments in the region in comparison with expansion of existing projects. In the process of analyzing this question we will look at the reasons for failure and success of projects.

The study is based on information obtained from personal interviews with FI and HC officials as well as review of available policy documentation and secondary data. Investor interviews provided a perspective on FI objectives and motivations, while interviews in the HC with government and trade association officials provided the HC perspective.[14] A comparison of HC and FI responses—including their similarities and differences—gave us a basis for evaluating the interaction process in historical perspective.

Since the large U.S. corporations account for most of the foreign investment in the forest-based sector, we concentrated our analysis on this group. This was further necessitated by the time and financial limitations of the study which prevented systematic sampling of the U.S. private companies and individuals with small investments in sawmills or other smaller, forest-based operations.

Appendix I lists the U.S. firms involved in Latin America. Some of these firms are strictly involved with paper conversion operations; that is, they import pulp and/or paper from outside the region and convert it to packaging materials, printing and writing papers, etc. Since our focus is on the forest resource being utilized, we eliminated such

[14] The basic sets of questions used in our interviews are reproduced as Appendix II.

firms from further detailed consideration.[15] In total, we identified forty-eight companies that had some seventy-five projects in the Latin American countries (excluding the Caribbean Islands).[16] In many cases, the companies only have one project and in others, only one U.S. project exists in a given country. Brazil, Mexico, and Colombia have the greatest number of projects.

Of the total number of companies involved, we contacted sixteen representing the major U.S. forest industry corporations involved in Latin America.[17] From response to the initial contact, we selected ten that involved primary processing of HC forest resources for a variety of products in several countries. These constituted the primary sample, and the companies were interviewed in the United States and in Latin America. In addition, we contacted some five other U.S. forest-based projects while we were interviewing in the HCs. Table 1 indicates the main characteristics of the projects included in the sample.

As noted in Table 1, several of the projects of the companies interviewed have failed. We felt that it was particularly relevant to include these projects in our analysis because of the high attrition rate in the forest-based sector. In addition, the sample includes three companies that have just expanded capacity of their Latin American operations or have operational plans to do so in the near future. Again, as explored in detail in Chapter Five, one of the facts that characterizes the present situation with regard to foreign investment in the sector in Latin America is the scarcity of new investments coupled with investment in expansion of existing projects.

Finally, in Chapter Six we review the potentials of the sector and consider implications of the study results, noting particularly those aspects related directly to exploitation of the forest resource. We make some suggestions concerning means of overcoming existing bottlenecks and conflicts.

Due to limitations of time and funds, the study is by necessity exploratory in nature and covers in detail only those questions which we felt were relevant to the objectives. In the context of the dynamic nature of the subject, the empirical conclusions should also be qualified as relating to particular conditions existing in 1971 and 1972, the years in which the empirical work was carried out.

---

[15] The list represents the situation as of early 1971. Some of the projects listed have since been abandoned, and others not included have come into being.

[16] We also identified some twenty-two projects which were fairly small and run or controlled by a single individual or small family companies in the United States.

[17] Several of those initially contacted had no involvement with primary forest-based activities and were thus eliminated from further consideration.

Table 1. Characteristics of Main Projects Studied

HC	Size of affiliate[a]	Type of product and market[b]	Forest resource information			Size of parent company[f]	No. of other affiliates in Latin America	Parent ownership (%)
			Type[c]	Control[d]	Area[e] (1,000 ha.)			
Colombia	Medium	PLY/VEN-DOM/EXP	TRP HDW	CONC	75+	Large	1	75
Brazil	Large	P+P-DOM	PLNT	OWN	(10)	Large	1	100
Brazil	Medium	VEN-EXP	TRP HDW	OWN	284	Large	1	100
Brazil	Large	ALL-EXP	PLNT	OWN	—	Large	0	100
Colombia	Large	P+P-DOM	TRP HDW	CONC	50	Large	2	Majority
Honduras	Large (failed)	P+P-EXP/DOM	CON NAT	CONC	1,495	Large	2	51
Colombia	Medium (failed)	VEN/PLY-EXP/DOM	TRP HDW	CONC	58	Large	0	95+
Honduras	Medium (failed)	PLY-EXP/DOM	TRP HDW	CONC	—	Large	1	Majority
Nicaragua	Small	PLY-EXP/DOM	TRP HDW (CON-NAT)	CONC	300+	Large	1	100
Honduras	Small	SWN-EXP/DOM	CON-NAT	MKT	(14)	Small	0	100
Ecuador	Large (failed)	VEN/SWN-EXP	TRP HDW	CONC	200–250	Large	1	100
Nicaragua	Medium (failed)	SWN/VEN-EXP	TRP HDW	CONC	—	Small	0	Majority
Brazil	Large	P+P-DOM	PLNT	OWN	32	Large	0	100

Notes:

[a] Based on total investment: Small, under $5 million; Medium, $5–20 million; Large, over $20 million.

[b] PLY, plywood; VEN, veneer; P+P, pulp/paper; SWN, sawnwood; EXP, export market; DOM, domestic market.

[c] TRP HDW, tropical hardwoods primarily; PLNT, plantation wood primarily; CON NAT, natural conifers primarily.

[d] OWN, land owned by affiliate; CONC, concessions from government; MKT, buys primarily market wood.

[e] Area controlled or owned by affiliate. 1 ha. = 2.47 acres. Parentheses indicate authors' estimate.

[f] This designation is an estimate based on the company's position relative to all other similar companies in the United States.

While these limitations need to be stated, we also believe that the results, conclusions, and suggestions might have much wider applicability for other sectors and possibly other regions. There is a great mass of the international investment literature that deals with generalities without reference to specific sectors or areas. The reader of such literature can often derive beneficial insights by deducing more specific implications for his particular interests. The present study deals as much as possible with specific projects, conditions, and countries, and we hope that the reader interested in other sectors and other areas or in foreign investment in general can derive as much value in terms of his interests by applying an inductive approach and his own experience with foreign investment in other sectors.

chapter two **Forests and Forest-Based Activity in Latin America**

This chapter provides an overview of the importance of forests and forest-based activities in Latin America relative to the rest of the world. Particular attention is paid to production, trade and consumption trends, and the problems and potentials associated with development of the sector.

Forest Resources of Latin America[1]

The total forest area of Latin America has been estimated at 900 million ha. This is about one-fifth of the total forest land of the world. More than one-third of this area is concentrated in Brazil (more than 350 million ha.). Nine percent is located in Mexico and Central America and the rest in South America. (Peru, Colombia, and Argentina each have around 70–80 million ha.) Of the forest area of 900 million ha., 530 million ha. are considered unproductive by the Food and Agriculture Organization (FAO) of the United Nations,[2] and another 20 million ha. are forest protection reserves. This leaves about 350 million ha. available for exploitation. In other words, Latin America has about 20 percent of the forest area of the world, but a

[1] Based on the findings of the FAO of the United Nations, *Latin American Timber Trends and Prospects;* B. Lamb "Tropical American Forest Resources"; and ECLA/FAO/UNIDO, Papers from the Regional Consultation on the Development of the Forest and Pulp and Paper Industries in Latin America.

[2] Unproductive Forests: "Forests where ecologically adverse conditions limit physical productivity to such an extent that all economic exploitation is impossible. . . . Several other types and subtypes of unproductive forest may be recognized, e.g., forest may be unproductive because of low economic productivity, where forest growth is too low to warrant industrial exploitation: other forests may produce a sufficient timber crop to warrant an industrial exploitation but transport costs to the nearest market may be prohibitive. In addition, several combinations of these factors may cause the forest to be unproductive" (United Nations, World Forest Inventory, May 1966).

higher proportion, equal to 36 percent, of the unproductive forests. This can be compared with the forests of Africa and Asia (excluding the USSR) which have 18.5 percent and 13.2 percent, respectively, of the forest area of the world and 27.3 percent and 10.9 percent of the unproductive forests. Of course, the concept of "unproductive" forest is a relative one, at least in the case of Latin America, where little information is available. To a great extent it depends on the general conditions of the available technology, infrastructure, and markets. For this reason, the area which is "unproductive" today can represent a major potential for future economic development.

Of the total forest area, at most 20 million ha., or 2.2 percent, are coniferous forest. While Latin America has an abundance of broad-leaved forests, it is critically short on coniferous wood. Recent estimates put the regional softwood raw material deficit in 1985 at about 24 million m^3, unless changes in technology are introduced which permit substantial substitution of broad-leaved species for conifers.

The most extensive type of forest in Latin America is the tropical rain forest (about half of the total forest area). This is a dense forest and heterogeneous in terms of species. This last characteristic is one of the main obstacles to its economic exploitation, since the valuable species often account for only a small proportion of the growing stock. Gross stocking is variable, but the average is around 200–300 m^3 per hectare, with only 10–30 m^3 per hectare or less being commercially usable at present.

In the dry deciduous forest, which covers an area of about 400 million ha., it is also possible to find some valuable timber species. However, the growth rate is very slow. Stocking is also low, ranging from 20–50 m^3 per hectare.

Another important element in the Latin American forest situation is the existence of significant areas of quick-growing, man-made forests. In 1964, they covered an approximate area of 1.6 million ha., or 37 percent, of the forest plantations of the less-developed world. Since then there has been a substantial increase. The main concentrations are in Brazil, Chile, Argentina, and Uruguay. Recent estimates place the total plantation area at some 2.2 million ha., with about 1.3 million ha. being broad-leaved (mainly eucalyptus) and 865,000 ha. being conifers.

It has been estimated that for eucalyptus pulpwood and pole plantations, the rate of growth is on the order of 20–25 m^3 per hectare per year on a six- to eight-year rotation, and for the quicker-growing pines and *Cupressus,* 12–17 m^3 per hectare per year on a slightly longer rotation. In Chile, the very fast-growing plantations of *Pinus radiata* have an average annual rate of growth of 24 m³ per hectare per year at the age of 26–30 years.[3] Rates over 35 m³ per hectare per year are by no means uncommon.

With these rates of growth, the fast-growing plantations can produce sawlogs within 20 years or even less, and they are obviously of interest to industry. On the other hand, in the natural tropical broad-leaved forest, the growth rates are very low—only 1–3 m³ per hectare per year, although these rates can be sharply increased to 10 m³ or more with appropriate silvicultural treatment.

The Latin American forests have an approximate growing stock of 78,000 million m³, or one-third of the total world growing stock, and they produced a volume of removals equal to 280 million cubic meters roundwood [m³(r)] in 1968, or 13 percent of the world total [and almost two-fifths of the total removals of the less-developed countries (LDCs)].

In the case of the native forests of Latin America, there is generally little knowledge of the wood properties and a lack of experience with utilization requirements. This is another reason why only a few species are normally usable for industrial purposes. But, since there are large areas with natural forest which could be exploited economically, and increasing technical knowledge with respect to wood properties can be expected, the native forest will continue to be a main source of forest products, although it is likely that man-made forests, either with native or introduced species, will be established on much of the exploited area. In the case of Chile, for example, more than half of the sawn-wood used in the country is obtained from plantations, whereas other important industries such as those of pulp and paper, and fiberboard, depend entirely on the *Pinus radiata* plantations. However, at present the natural forests of Latin America represent the largest reserve of forest resources in the Western Hemisphere. It is likely that this reserve will increasingly supply forest products to the developed world.[4]

With respect to ownership of forests, in Central America 45 percent of the forest land is public as compared with 56 percent in South America. These figures are below the world average of 77 percent for publicly owned lands. In Brazil and Peru—the two countries that have the most forest land in Latin America—the public forest lands are 43 percent and 94 percent of the total, respectively.

[3] Government of Chile, "Inventario de las Plantaciones de la Zona Centro Sur de Chile."
[4] S. Pringle, in "World Supply and Demand of Hardwoods," Appendix C, p. 44, suggests that Latin America's contribution to tropical broad-leaved woods exports might increase from 3 percent in 1967 to about 10 percent in 1985, or from a log equivalent of 1.1 million cubic meters roundwood [m³ (r)] to 6.4 million m³ (r).

The largest untapped forest resource in the world is found in the Amazon Basin. The problems in utilizing this vast resource are substantial, and this accounts for the lack of major developments to date. Other untapped forest areas are found in the Peten of Guatemala, the Atlantic coast of Honduras, Nicaragua, Costa Rica, southwestern Panama, Mexico, Venezuela, and the Guiana uplands, and the Pacific uplands costal forest of Panama, Colombia, and Ecuador.

Forest Industries and Production[5]

There is no adequate estimate of the number of sawmills in Latin America. However, the number is well above 15,000, when small, intermittently operating units are included. In general, the industry is characterized by a great number of small, inefficient units and a few large, well-equipped mills that produce for both export and domestic markets. The major problems to be overcome relate to the quality of the product, the high sales prices and lack of sales standards, poor distribution and transport systems, discontinuities of log supply, and the lack of adequate treatment of the product.

As of 1967, there were some 234 mills producing wood-based panels in the region, with an estimated overall production capacity of some 1.2 million tons per year. Plywood accounted for roughly half of the total, with fiberboard and particleboard accounting for the remaining 50 percent. Four countries—Argentina, Brazil, Mexico, and Venezuela—accounted for about 70 percent of the total production. The region had an export surplus for this product group of about $1 million in 1969, which is not very large, considering the size of the region and its forest resource base.[6]

The pulp and paper industry in the region is again characterized by a great number of mills (some 640 in 1967), with only a few of them producing the bulk of the output. Only twenty-six of the total number of mills had annual capacities exceeding 60,000 metric tons. Few of the mills are integrated pulp and paper production units, and most of the smaller units are operating with old, inefficient equipment. Costs

---

[5] The most detailed recent discussion on forest industries in Latin America is to be found in the various papers prepared for a United Nations Regional Consultation on the Development of the Forest and Pulp and Paper Industries in Latin America, Mexico, D.F., May 19–26, 1970.

[6] United Nations, Economic and Social Council, Economic Commission for Latin America, *Account of Proceedings and Recommendations of the Regional Consultation on the Development of the Forest Pulp and Paper Industries of Latin America,* p. 13.

are very high, except in the small number of large, efficient mills, many of which are affiliates of U.S. and other foreign corporations. The two products of particular importance in terms of pulp and paper development in the region are newsprint (with more than $100 million net imports in 1970) and packaging paper (which accounts for about 54 percent of paper consumption in the twenty-two major Latin American countries).[7]

Latin America increased its production of forest products at rates above the average for the world and above the average of the developed market economies during 1966–1970, except in fiberboard, particleboard, and newsprint. The most important increases were 42.9 percent in plywood, 40 percent in other paper and paperboard, and 30 percent in printing and writing paper. As a consequence of these rates of growth, the participation of Latin America in the total world production of most forest products has increased slightly between 1966–1970 (Table 2).

Trade in Forest Products

Latin America is not a very important region of the world if we look at its share of total exports or imports of forest products, but it is important at least in some subsectors if we take the developing market economies as a point of reference (Tables 3 and 4). This is true in the case of pulpwood, sawnwood, fiberboard, wood pulp, and newsprint. Its importance as an importer in the context of the developing world is rather low except in newsprint and wood pulp. In terms of total value of net imports, the position of the region has suffered a substantial deterioration from about $300 million in 1966 to $400 million in 1970. Major importing countries are Argentina, Mexico, and Venezuela. Only Chile and Brazil showed a large export surplus in 1970 (Table 5).

Market Size

The size of the Latin American market for forest products can be approximately measured by the level of apparent consumption.[8] Latin American consumption trends since 1959 are shown in Table 6. Even

[7] Ibid., pp. 18–19.
[8] Apparent consumption = production + imports − exports.

Table 2. World Production of Forest Products, 1959, 1966, and 1970

	Industrial wood (million m³ (r))[a]	Sawnwood (1,000 m³ (s))[b]	Plywood (1,000 m³)	Particle-board (1,000 metric tons)	Fiberboard (1,000 metric tons)	Newsprint (1,000 metric tons)	Printing and writing paper (1,000 metric tons)	Other paper and paperboard (1,000 metric tons)	Wood pulp (1,000 metric tons)
Total world									
1959	1,009	335,807	14,758	1,346	4,109	13,018	12,995	42,892	54,905
1966	1,153	375,666	25,348	6,226	6,312	18,289	20,953	65,295	85,130
1970	1,275	407,507	32,607	11,782	7,778	21,392	26,476	78,872	102,570
Developed market economies									
1959	563.6	178,809	12,086	1,156	3,636	11,897	11,279	35,885	49,115
1966	662.7	204,334	20,805	4,685	4,857	16,302	18,191	54,150	75,413
1970	719.0	219,645	26,386	9,017	5,746	18,899	23,040	64,379	90,224
Developing market economies									
1959	90.5	21,602	763	41	135	205	551	1,512	583
1966	126.2	30,718	1,941	212	270	429	1,189	2,681	1,719
1970	156.7	38,142	3,126	538	360	516	1,485	4,053	2,320
Latin America									
1959	34.6	11,471	288	19	88	139	272	1,080	490
1966	38.8	13,242	399	121	147	265	502	1,904	1,287
1970	47.6	15,898	570	383	171	288	654	2,670	1,664
Percentage of word total									
1959	3.4	3.4	1.9	1.4	2.1	1.1	2.1	2.5	0.9
1966	3.4	3.5	1.6	1.9	2.3	1.4	2.4	2.9	1.5
1970	3.7	3.9	1.7	3.2	2.2	1.3	2.5	3.4	1.6

Source: United Nations, Food and Agriculture Organization, Yearbook of Forest Products. Several issues.
[a] Million m³ (r) = million cubic meters roundwood.
[b] 1,000 m³ (s) = 1,000 m³ processed product.

Table 3. Exports of Major Forest Products, 1966–1970

(1,000 U.S. dollars)

	Pulpwood	Logs	Sawnwood	Wood-based panels	Newsprint	Printing and writing paper	Other paper and paperboard	Wood pulp
World								
1966	160,384	805,048	2,022,404	744,763	1,235,191	458,558	1,305,836	1,660,623
1970	223,421	1,445,758	2,627,591	1,144,068	1,491,710	850,570	1,962,003	2,493,731
Developed market economies								
1966	71,433	213,371	1,297,087	524,784	1,189,489	432,925	1,244,630	1,578,965
1970	109,407	489,161	1,773,733	726,410	1,434,959	813,690	1,839,150	2,369,997
Developing market economies								
1966	4,673	493,296	226,282	147,714	10,082	10,876	12,347	29,662
1970	6,584	776,347	33,631	323,529	10,743	10,261	21,385	46,207
Latin America								
1966	4,041	19,122	88,260	14,925	1,932	639	2,926	14,024
1970	6,044	12,080	117,416	31,302	9,595	4,151	7,727	21,014
Percentage of world total								
1966	2.5	2.4	4.4	2.0	0.2	0.1	0.2	0.8
1970	2.7	0.8	4.5	2.7	0.6	0.5	0.4	0.8

Source: United Nations, Food and Agriculture Organization, *Yearbook of Forest Products.* Several issues.

Table 4. Imports of Major Forest Products, 1966–1970

(1,000 U.S. dollars)

	Pulpwood	Logs	Sawnwood	Wood-based panels	Newsprint	Printing and writing paper	Other paper and paperboard	Wood pulp
World								
1966	187,430	1,226,699	2,275,291	777,736	1,393,411	390,440	1,445,763	1,807,337
1970	285,529	1,206,134	2,979,816	1,180,396	1,614,288	744,997	2,135,874	2,632,641
Developed market economies								
1966	170,527	1,047,157	1,911,067	666,333	1,202,228	278,055	1,010,168	1,576,478
1970	266,441	1,909,948	2,536,304	1,019,632	1,349,001	505,359	1,483,985	2,257,789
Developing market economies								
1966	1,161	113,157	224,813	72,505	176,464	103,415	307,634	132,128
1970	2,353	244,391	291,310	89,729	231,498	190,490	437,614	206,463
Latin America								
1966	—	11,964	84,948	13,125	103,941	18,823	112,665	80,400
1970	—	14,530	114,464	15,770	140,416	49,946	140,745	114,326
Percentage of world total								
1966	—	1.0	3.7	1.7	7.4	4.8	7.6	4.4
1970	—	1.2	3.8	1.3	8.7	6.7	6.6	4.3

Source: United Nations, Food and Agriculture Organization, *Yearbook of Forest Products.* Several issues.

Table 5. Total Net Imports (−) and Net Exports (+) of Major Forest Products
in Selected Latin American Countries, 1959, 1966, and 1970

(1,000 U.S. dollars)

Country	1959	1966	1970
Latin America	−179,702	−294,401	−400,252
Argentina	−70,211	−119,323	−157,163
Bolivia	−622	−1,365	−476
Brazil	−9,441	+43,372	+39,661
Chile	+4,085	+16,311	+30,388
Colombia	−7,377	−24,734	−17,932
Ecuador	−2,815	−944	−1,807
Guiana	+296	−326	−1,346
Paraguay	—	+670	+2,792
Peru	−9,080	−15,779	−16,471
Surinam	+430	+2,282	+2,226
Uruguay	−10,825	−11,265	−12,303
Venezuela	−24,664	−31,341	−49,941
British Honduras	+2,170	+1,263	+313
Costa Rica	−3,156	−8,097	−17,034
Cuba	—	−29,255	−31,000
Dominican Republic	−3,201	−1,784	−4,570
El Salvador	−2,352	−10,617	−9,602
Guatemala	−655	−3,206	−6,392
Haiti	—	−398	−1,236
Honduras	+4,406	−5,568	−2,999
Mexico	−24,844	−43,838	−85,805
Nicaragua	—	−134	−36
Panama	−2,171	−11,412	−13,770

Source: United Nations, Food and Agriculture Organization, *Yearbook of Forest Products.* Several issues.

Table 6. Apparent Consumption of Forest Products in All Major Latin American
Countries, 1959, 1966, and 1970

Product	1959	1966	1970
Sawnwood (1,000 m³)	11,366	12,863	15,379
Plywood (1,000 m³)	288	414	585
Particleboard (1,000 metric tons)	16	111	371
Fiberboard (1,000 metric tons)	101	139	137
Wood pulp (1,000 metric tons)	910	1,828	1,693
Paper and paperboard (1,000 metric tons)	2,326	3,831	5,243

Source: United Nations, Food and Agriculture Organization, *Yearbook of Forest Products.* Several issues.
Note: Apparent consumption = production + imports − exports.

though the level of apparent consumption in Latin America is only a
fraction of that of the United States for most products, the region
accounts for about half of the consumption by all LDCs put together
(Table 7). For most products, the rate of increase in the level of con-

Table 7. Apparent Consumption of Forest Products in the United States and All Less-Developed Countries, 1970

Product	United States	All LDCs
Sawnwood (1,000 m³)	95,978	37,256
Plywood (1,000 m³)	15,775	1,317
Particleboard (1,000 metric tons)	2,024	548
Fiberboard (1,000 metric tons)	2,532	429
Wood pulp (1,000 metric tons)	37,704	3,158
Paper and paperboard (1,000 metric tons)	49,152	9,941

Source: United Nations, Food and Agriculture Organization, *1969–70 Yearbook of Forest Products.*
Note: Apparent consumption = production + imports − exports.

Table 8. Apparent Consumption of Forest Products in Four Major Latin American Countries, 1970

Product	Argen-tina	Brazil	Colom-bia	Mexico	Total
Sawnwood (1,000 m³)	1,351	6,961	1,605	1,522	11,439
Plywood (1,000 m³)	120	141	65	98	424
Particleboard (1,000 metric tons)	78	176	14	36	304
Fiberboard (1,000 metric tons)	31	44	7	19	101
Wood pulp (1,000 metric tons)	306	750	106	489	1,651
Paper and paperboard (1,000 metric tons)	911	1,167	284	1,262	3,624

Source: United Nations, Food and Agriculture Organization, *1969–70 Yearbook of Forest Products.*
Note: Apparent consumption = production + imports − exports.

sumption is quite high. For example, in the last five years, consumption of sawnwood rose by 20 percent, whereas in the case of most more industrialized products such as paper and paperboard the increase was nearly 40 percent. The rapid growth of demand explains in part the sharp increases in the import bills that Latin American countries have been facing during the last years.

Another important characteristic of the Latin American market for forest products is that consumption is heavily concentrated in four countries, which together account for about three-fourths of the Latin American market. These countries are Argentina, Brazil, Colombia, and Mexico (Table 8).

We can summarize the overall situation in Latin America as follows:

1. The forest resource of the region is substantial.
2. By far the largest portion of Latin American forest resources are

in the broad-leaved category, including nearly 1.5 million ha. of broad-leaved, man-made forests.

3. There is a shortage of coniferous or softwood resources in the region, although some countries, such as Honduras and Chile, have substantial coniferous resources.

4. In general, the region is not expanding rapidly in absolute terms in forest-based exports, although the potential for future expansion is there if modern technology, management, and marketing expertise are introduced.

5. There are still a number of sizeable areas of forest which could offer attractive potentials for development (the situation differs widely by countries).

6. Much of the recent activity in the region has been in man-made forests.

7. The region has a significant and sizeable net import bill in forest products—primarily in the pulp and paper category—a situation which could be reversed with better management of the resource.

8. The region has a sizeable and growing internal market for forest products.

9. This market is almost completely concentrated in Argentina, Brazil, Colombia, and Mexico.

The general lack of local experience in developing modern, efficient forest industries and the possibilities for reducing the import of forest products makes foreign involvement in the sector attractive to many of the countries in the region. Moreover, the resource potential of the region and the growth of both internal and export markets are such that, if coupled with various policy reforms and developments in transportation and marketing infrastructure, they could make Latin America an increasingly attractive area for foreign investment. Thus, there appears to be a broad area of mutual interest between HCs in the region and potential FIs in forest industries. Problems in defining and pursuing these interests to mutual advantage are discussed in the chapters that follow.

chapter three **Host Country and Foreign Investor Motivations**

There are a number of different objectives and expectations of foreign investors (FI) and host countries (HC) which motivate them in their involvement with foreign investment projects and help explain the interaction between the HC and FI. This chapter summarizes the objectives of Latin American countries and FIs in the context of investment in the forest-based sector.

Host Country Objectives and Motivations

Why should a country want or encourage foreign involvement in the development of its forest resources? The main benefits that foreign investment can bring are increased availability of capital, foreign exchange, managerial and technical skills, and access to new markets. A foreign investment project also provides employment, value added, and tax revenues the same as domestic investments.[1] Of course, all these contributions are to a great extent "potential" in any given case. Also, many of them are accompanied by negative aspects for the country, which, as we will discuss later, provide the grounds for the emergence of conflicts.

The inflow and outflow of foreign exchange resulting from foreign investment is often of main concern to the HC. As a first approximation, we can consider the net foreign exchange earnings as being determined by the following: $E = X + Ms + Ig - D - M$.
In which:

E = Net foreign exchange earnings due to investment.
X = Exports generated by foreign investment.

---

[1] On the contribution of foreign investment, see R. F. Mikesell, ed., *Foreign Investment in the Petroleum and Mineral Industries: Case Studies of Investor-Host Country Relations,* particularly chapters 1 and 16.

23

M = Imports generated by foreign investment.
Ms = Imports substituted by foreign investment.
Ig = Gross investment from foreign sources.
D = Dividends, depreciation, interests, etc. transferred abroad.

One concern of the HC, with respect to the foreign exchange impact of foreign investments, is the size of E.

Another contribution of foreign investment is its addition to the real capital resources of the HC. In this case, the benefits not only consist of the inflow of capital but also the inducement to mobilization of domestic capital in a sector. Both can generate increased income, which is the real objective of the country.

The productivity of capital and other resources is seriously impaired if there is a lack of other complementary inputs, of which managerial and technical skills tend to be critically scarce in the LDCs. Foreign investment is one source of these types of skills. If investment does not provide new technology immediately, research and training initiated by a foreign enterprise can mean significant future contributions.

However, even if new improved technology is introduced, the actual contribution to the HC could be very different from the potential benefits if the results do not permeate to the rest of the economy, that is, if the investment merely creates a technological enclave. The extent of induced innovation, the impact of the introduction of new technology already available abroad, and the efficiency of the actual diffusion process are aspects that involve serious problems of measurement.[2] Basic technology can be obtained without becoming involved with foreign enterprises. However, the transference and adaptation of advanced technology that is essential to remain competitive in international markets often requires the active participation of foreign enterprises.[3]

Finally, we have already mentioned that LDCs see the growing need for establishing competitive new export industries, and in order to do this it is necessary to have good market connections. Such connections can often be provided most effectively by an international company which, in many cases, provides a captive market itself and has a better bargaining position in other markets, more information, and better marketing facilities.[4]

[2] Y. Hayami and V. Ruttan, *Agricultural Development: An International Perspective.*

[3] Cf. United Nations Conference on Trade and Development, *Establishment of Tropical Timber Bureaux,* p. 30.

[4] This is particularly important in the case of new tropical hardwood products such as moldings, veneers, and many others. This has been stressed by both local entrepreneurs and marketing experts in the field.

While these are the main potential contributions, the question still remains as to what motivates a country to accept or encourage direct foreign private investment rather than seeking some other source of the above benefits such as international technical and financial aid.

A simple answer which applies in some cases is that the country has the type of political and economic structure which gives preference to private investment, whether it be foreign or domestic, over government investment, or accepts both equally as sources of capital and expertise needed for development. Such countries, rightly or not, have little fear of foreign domination and political control through foreign investment.

Another simple, somewhat related answer is that in some cases the ruling elite in the HC see foreign private investment as a source of personal gain, a gain which would be more difficult to obtain if the capital and technical aid came through international organizations such as the United Nations or a regional development bank. The reasoning is that private FIs are willing to "share the pie" if it is large enough to insure them an acceptable profit or some other strategic value which they seek.[5]

Finally, since direct foreign investment and other forms of support are not mutually exclusive, foreign firms could be sought as a complement to other alternatives, especially when the country is already over-committed with other sources. Thus in some circumstances the foreign firm could be the best or only source of additional technical know-how, foreign exchange, capital markets, etc. not available domestically.

In all cases, regardless of the political aspects, the necessary resources to develop particular sectors or projects are not forthcoming from domestic or other international sources, either because they are not available or because they are being channeled into other sectors or projects which appear more favorable from a domestic point of view. In the latter case, it is not necessarily profitability alone which determines the more favorable alternatives. The choice might be related to perceived risks, social and cultural biases, market-access problems, etc.—all of which have been important considerations in the forest-based sector of various countries.

So far we have discussed the potential contributions of foreign investment to the fulfillment of the HC's objectives. But, is there any intrinsic advantage in promoting foreign investment in the forest-based sector?

The opportunity costs involved in natural tropical forest exploitation

[5] Several persons interviewed stressed the importance of personal gain as a motive.

are relatively low for most Latin American countries since there are few alternative demands for such resources and since the net value growth of the mature natural forest is close to zero (that is, growth potential is not being destroyed). This means that the impact on development could be greater than in other sectors and that the motivation for encouraging investment can also be greater than in the case of mineral or petroleum resources; for example, where the opportunity cost of present use is relatively higher in terms of future use, since such resources do not deteriorate in the ground, like wood resources do on the surface.

The parallel argument is often made that there is an opportunity cost involved in *putting off* development in the forest sector or in not utilizing part of the wood being lost. There are two aspects to this argument. First, while the natural tropical forest is in balance (that is, there is a zero net growth), such forest land has the potential to produce 10 m³ of wood or more per year per hectare if properly utilized and managed. Depending upon economic conditions (markets, production costs, etc.), loss of this potential may represent an opportunity cost to the country. The second argument is that the total forest base (the stock) is being significantly depleted by shifting cultivation which produces few, if any, benefits from the timber removed.[6] Where there is no market for the timber destroyed the opportunity cost would be zero. However, there may be potential markets available that could absorb the wood at a price high enough to cover costs and leave a net return to the settlers (or the government). In some of these latter cases, the capital and technical know-how plus markets needed to efficiently utilize the timber resources are not available domestically, whereas they may be available overseas with the help of FIs.

This again contrasts with minerals and petroleum, when a delay in exploitation does not involve destruction of the resources and consequently no loss in the sense that a loss of forest capital occurs through decay, reduced growth, and uneconomic forest destruction. Increasingly, governments are recognizing this, and a stronger emphasis is being placed on forest resource development in several Latin American countries.[7]

Are these considerations important to the countries of Latin America in their decisions affecting foreign investment in the forest-based sector? If they are, which of them are most important? To find answers to these questions two approaches were followed. The first

[6] FAO estimates a loss of 5–10 million ha. per year.
[7] For example, in the case of the Brazilian and Peruvian Amazon settlement programs and in Colombian land use development policies.

Table 9. Reasons for Host Country Acceptance of Foreign Investment in the Forest-Based Sector

Reason	No. of responses for each reason[a]
Provides capital	4
Related to technology	6
Access to new technology	3
Increased availability of human capital	2
More training for domestic technicians	1
Increased international market opportunities (balance of payments)	6
Access to new markets	4
More value added to exports	2
Increased employment	4

Source: Based on personal interviews with eleven HC officials in five countries—Honduras, Nicaragua, Guatemala, Colombia, and Brazil.
[a] Number of interviewees who considered objective as first or second most important.

involved a survey of opinions among government officials having the responsibility of accepting or rejecting foreign investment applications in the forest-based sector. The second and complementary approach consisted of a detailed analysis of the regulations and statements contained in the foreign investment and forest legislation of the HCs.

Table 9 indicates the various motivations for attracting foreign investment in the forest-based sector given by government officials interviewed in the case study countries. As indicated in Table 9, the two most frequently mentioned motivations relate to the transfer of technology and managerial skills and access to export markets. The latter is obviously related to the basic need in some countries to increase foreign exchange availability through exports or import substitution. The former is related to a complex of factors, including some that are directly tied to the lack of local experience and technology associated with efficient forest exploitation for commercial purposes.[8]

The motivations and objectives expressed in Latin American foreign investment and forestry legislation are associated with:

1. A desire to improve the foreign exchange position as expressed, for example, through restrictions on capital repatriation, profit remittance disincentives, and export incentives.

2. A desire to accumulate increased capital which can employ addi-

[8] Related to this, an official of a forest industry association in Colombia suggested that Colombians who have capital are more prone to invest it in the cities, in such well-known activities as textiles rather than in the relatively unknown tropical forest hinterlands, where living conditions are not ideal, uncertainty is high, etc. Further, domestic firms do not want to pay high enough salaries to get competent people to live in the forest regions.

tional national citizens (as expressed in restrictions on local borrowing, fiscal incentives to attract capital, regulations with regard to minimum permissible domestic labor content in a project, local value-added requirements, etc.).

3. A desire to have control over national natural resources (as expressed in provisions setting limits to foreign ownership of resources, incentives for local participation in projects, etc.).

4. A desire to have technology transfers that can be freely adapted in the countries (as expressed in regulations regarding technology transfers, patents, rights, and royalties).

5. A desire to build some permanence into forest-based industries (for example, as expressed in conservation and sustained-yield provisions of forest laws).

Legislation identifies motivations for encouraging foreign investment —for example, increased employment, increased capital, improvement in foreign exchange position, and technology transfers—but at the same time it provides guidelines with regard to the "minimum acceptable" conditions for foreign investment. As such, fundamental conditions and attitudes which the foreign investor faces in the countries are often expressed in legislation. In addition, however, forestry laws often provide some specific guidelines for investment in the forest-based sector.

For example, in the area of national control and ownership of natural resources, foreigners entering into forestry investment projects in Latin American countries find that they often cannot obtain ownership rights to the forest lands on which they will be relying. Instead, the common type of arrangement is a timber concession agreement with the government.[9] The range of specific policies toward foreign investment in forestry is wide. In some cases, notably Mexico and now Honduras, it is specifically provided that only persons or companies from the HCs may operate forests commercially. In other places, for example, Guatemala, national companies or persons are given preference over foreigners in forest exploitation.

Review of the relevant legislation on forestry and foreign investment and the results of our interviews with HC officials indicate that the objectives of different countries are similar and that there is a basic

---

[9] Some U.S. investors own sizeable areas of forest land in certain Latin American countries, for example, in Brazil, but this is the exception and it is being discouraged to an increasing extent in most countries.

consistency between formally expressed objectives in legislation and those voiced by the government officials interviewed.[10]

Finally, it should be mentioned that laws and regulations play an important role in determining the type of investment project which will be undertaken in the forest-based sector. For example, a growing number of countries are prohibiting log exports while they encourage investment in the processing industries. In Latin America, most countries have either forbidden log exports by law or have made it extremely difficult to export them.

Foreign Investor Objectives and Motivations

There is no one accepted theory of investment behavior, particularly none relating to foreign investment. One major reason for this is the lack of an overall answer to the question of the motivations behind the decision to invest.[11]

However, given the desire or motivation to invest, the main question of interest here relates to the reasons why forest industries invest in particular regions and locations. The work done in location theory and the empirical studies dealing with forest industry location are of relevance to this question.

The basic assumption underlying a location analysis is that the firm has decided to invest. The question is then asked, Where should the investment be made? In terms of the present study, we are essentially asking, Why did or why should the investor want to locate his investment in a given location in Latin America?

This question helps to focus on the point that Latin America is not a homogenous region. The differences between countries, and regions within countries, are in many cases greater than those between areas such as southern Brazil and parts of the United States. We are dealing with a continuum of conditions and opportunities, not just two regions in the developed and the less-developed worlds. As such, it would be fruitless to search for generalities separating Latin America as a whole from the rest of the world in terms of investment decisions.

The approach used in most location analyses of forest industries is

---

[10] A divergence could be expected, especially when a new government with a different orientation toward foreign investment takes office. In the period needed to change the old legislation, a difference could exist.

[11] On this point, cf. the study of Y. Aharoni, *The Foreign Investment Decision Process.*

to assume that cost minimization is the dominant objective, with certain other considerations acting as constraints.[12] In this context, the cost factors which appear to be of major importance are wood, transportation (location of markets), labor cost and availability, financial assistance and taxes. The ordering of these various factors depends on the subgroup of the industry being considered (for example, sawmilling, particleboard, plywood, veneer, pulp) and the spatial "breadth" of the decision (for example, choosing a region, community, or specific plant site).

Hagenstein found in the case of lumber and wood pulp industry location in the northern Appalachians that the most important factor was cost of wood, followed by transportation (which relates to market proximity), and then labor. In the case of wood pulp, two other items were important—electricity and state and local taxes. But in both cases, wood, transportation, and labor dominated. The other industry group he looked at was particleboard, and in that case the cost of labor was most important, followed closely by wood, and then transportation.[13]

Schuster and Pendleton found the following factors to be important in the selection of a region in which to locate a veneer mill: (1) location of markets, (2) location of domestic raw materials, (3) transportation cost of imported wood, (4) costs of financial assistance, and (5) personal preference.[14] Other studies have found the same factors to be important, with wood, labor, and transportation (market location) tending to be the most critical ones.[15]

Based on our interviews with U.S. corporations having projects in Latin America, most of the same factors are also the major ones influencing investment decisions in particular countries and locations.

Most of the companies interviewed pointed out certain minimum requirements or prerequisites for investment which they have firmly in mind during their search for new opportunities. These (minimum requirements) include (1) acceptable net returns (either overall or through separate profit center), (2) HC stability (generally defined somewhat differently by each company), (3) tax and other laws which do not overly restrict the parent company as well as its affiliate

[12] Cf. Ch. H. Wolf, *Wood Industry Location Decisions*.
[13] P. Hagenstein, "Factors Affecting the Location of Wood-Using Plants in the Northern Appalacians."
[14] E. G. Schuster and T. H. Pendleton, "Decisions on Locating a Veneer Plant," as discussed in Ch. H. Wolf, *Wood Industry Location Decisions*.
[15] Wolf, ibid., provides a good bibliography of the work that has been done on forest industries' location decisions.

operations, and (4) types of contracts which firmly establish the legal basis for projects.

A common reason for investment given by respondents outside the pulp and paper subsector was to obtain and/or expand a supply of raw materials or semifinished products for the parent company or for other affiliates of the same company. These motivations relate to company growth goals as well as to stability or security objectives. Sometimes the motivation is based on a desire to secure a reserve of raw materials for the future. Such a reserve permits firms to increase levels of production when market conditions appear to be especially favorable, thus giving additional flexibility to the firm. The motives are primarily precautionary and speculative.

Most of the companies involved with pulp and paper indicated that their main objective was to obtain a share in a potentially attractive market or to protect an already established market.[16] These motivations relate to company expansion objectives.

The desire to increase profits was given as the *primary* inducement by one corporation.[17] Its subsidiary is not producing in order to supply other branches of the company, and it acts entirely as a separate profit center. For the other companies, an acceptable level of profits was important, but beyond such a minimum, profitability alone was not the major direct criterion for *particular* foreign investment projects.

Some forest-based investment choices overseas have been made on the basis of personal decisions of chief corporate officers with little analysis of alternatives. Tradition or familiarity with given areas were cited as important determinants of particular investments. This type of motivation and choice process is discussed in detail elsewhere.[18]

In the past, detailed analysis of alternatives was apparently uncommon for new investments.[19] In contrast, the decision to expand, even in relatively independent subsidiaries, was analyzed more carefully and systematically by the parent company. Part of the problem with analysis of new ventures is the poor information available on

---

[16] The former motive is also related to overcoming HC import barriers. Both are also related to transport cost considerations.

[17] However, for a number of companies, their foreign involvement is motivated by U.S. tax advantages which affect overall profits. See J. N. Behrman, "Promotion of Private Foreign Investment," p. 188, for relationships to tax advantages as a motivation.

[18] Cf. Aharoni, *The Foreign Investment Decision.*

[19] This is supported in general by J. Dean, "Measuring the Productivity of Capital"; and J. B. Mathews, "How to Administer Capital Spending." See also W. Mead, "Long-Term Investment Planning for Forestry Development."

specific areas and conditions.[20] On the other hand, the established subsidiary produces a flow of information that can be used in comparing performances among subsidiaries of the parent company as well as in making decisions with regard to expansion investments.

Our findings in this respect coincide with results obtained for sectors other than forestry. For example, Piper[21] studied twenty-one enterprises operating in Latin America, producing a variety of products such as fertilizers, processed vegetables, corn, beef, etc., and found that a multicountry comparison was rarely made and that, in general, U.S. firms, large and small, approach foreign investment projects with much less sophistication than in the case of domestic projects. The implication for those countries that seek an increased flow of foreign investment directed to their forest-based sector is that direct government contact with firms that are considering the possibility of investing abroad is important in order to encourage the creation of new ventures. On the other hand, if a subsidiary becomes established and successful, it is likely that the parent company will favor expansion of the existing enterprise over new investment in a different country, other things being equal. This assertion is based on one of our survey results showing that nearly half of the sample companies have recently expanded capacity of existing projects or have made commitments to do so in the near future, while only one company (as of 1971) had made a commitment for new investment in a country other than the ones in which it is currently operating. Some of the reasons for this bias will be discussed in Chapter Five.

[20] Cf. D. DeBoer, "Impressions of Industrial Problems in Tropical Forestry," on the impact of inadequate information.
[21] J. Piper, "How U.S. Firms Evaluate Foreign Investment Opportunities."

chapter four **U.S. Direct Private Investment**

Much of the modern part of the forest-based sector in Latin America is controlled by foreign corporations. This generality holds most nearly for the more technically complex and capital-demanding activities such as pulp and paper production, and least for such activities as sawmilling.[1]

In the case of pulp and paper, the percentage of U.S. control in Latin America has been quite substantial. In Brazil, which is the largest producer of pulp and paper in Latin America and also the largest recipient of U.S. direct private investment in the forest-based sector, U.S. affiliates had sales of some $47 million in 1966, and accounted for a little more than one-third of total Brazilian sales.[2] Total sales of U.S. manufacturing affiliates in paper and allied products in Latin America amounted to some $292 million in 1966.[3] Total Latin American sales of paper and paperboard in that year are estimated to have been between $500–600 million. U.S. affiliates therefore accounted for around 50 percent of the total.[4]

U.S. involvement in sawmilling in Latin America is slight according to U.S. Department of Commerce statistics, which, however, only consider enterprises with employment of 100 persons or more. In the

[1] The discussion presented in this chapter is based on Th. J. Lambiase, *U.S. Direct Investment Abroad in Paper and Allied Products;* H. K. May, *The Effects of the United States and Other Foreign Investment in Latin America;* U.S. Department of Commerce, "Paper Industry Leads in Foreign Growth," and *U.S. Direct Investment Abroad–1966,* and information gathered in our interviews with companies and host country (HC) governments.

[2] The year 1966 is used as the point of reference in this discussion since it is the latest year for which complete data on U.S. involvement are available. (See U.S. Department of Commerce, ibid.).

[3] May, *The Effects of the United States,* Table 3.

[4] The figure for U.S. affiliates includes paper conversion. An allowance for conversion activity is included in the total production figures for Latin America. It is difficult to derive an adequate average value for production of paper and paperboard. The figure of $200 per metric ton is thought to be representative of the order of magnitude and was used in this study, but the resulting total value figure should be treated with caution and as an indication of the order of magnitude only.

wood-based panel industry, there is also less foreign involvement than in pulp and paper.[5] However, even in a large lumber-producing and -exporting country such as Honduras, it has been estimated that some 60 percent of total capital in sawmilling and wood-based panels was in foreign hands as of 1972. Total sales of U.S. affiliates in the lumber and wood products industries in Latin America amounted only to about $13.5 million in 1966.[6]

As a point of comparison, in the Canadian forest-based industries some 41 percent of total sales in the paper and allied products industries were controlled by majority-owned foreign (mainly U.S.) companies in 1967. For the total forest-based sector, the comparable figure is estimated to be 33 percent. In other words, U.S. influence or control in the sector in Canada is quite similar in proportion to total activity as in the case of U.S. investment in Latin America.

Total U.S. direct foreign investment (fixed-asset expenditures) in the paper and allied products sector was $312 million in 1966, with Latin America receiving about 10 percent, or $30 million, of this total. This compares with about $79 million total investment in 1960, with about 7–8 percent, or $5–6 million, in Latin America. Total asset value of foreign affiliates in this sector at the end of 1966 was $1,508 million, with slightly more than 10 percent, or $150–160 million, in Latin America.

While current U.S. investment and total fixed assets in the paper and allied products sector in Latin America each comprised only about 10 percent of the world totals in 1966, net earnings from Latin America in that year were about $27 million, or roughly 21 percent, of the total net income from U.S. foreign investment in the sector.[7] In other words, U.S.-owned Latin American operations provided greater net earnings per dollar of investment or fixed assets than did operations in other foreign areas. About $18 million, or roughly 70 percent of

---

[5] There are exceptions. The Dutch firm of Bruynzeel, N.V., for example, accounted for some 40 percent of total Latin American production of particleboard in 1966.

[6] This should be considered a very conservative figure. It is based on a study by May, *The Effects of United States,* which indicates that only about 75 percent of all sales by U.S. manufacturing affiliates was included in the basic statistics provided by the U.S. Department of Commerce for firms with more than 50 percent U.S. ownership and with employees numbering more than 100 persons. Many of the firms in the lumber and wood products category do not fall within this definition, so probably the percentage of sales not reported in the U.S. Department of Commerce statistics is even higher for this sector, which is characterized by small and medium-sized units.

[7] Lambiase, *U.S. Direct Investment Abroad.* The figures given do not reflect adjustments for U.S. tax purposes, nor are they adjusted for changes in international values of currencies.

Table 10. Local Expenditures of U.S. Majority-Owned, Forest-Based Affiliates in Latin America, 1966

($1 million U.S.)

Item	Paper and allied products	Lumber and other wood products
Materials and supplies	128	6
Payroll	33	3.5
Wages	13	(2)
Salaries	11	(1)
Social benefits	9	(Less than $0.5 million)
Interest	3	Less than $0.5 million
Taxes	13	1
Other	5	1
TOTAL	182	11

Source: May, *The Effects of the United States.*

net earnings in 1966 were reinvested in Latin America. Net capital flow (that is, in addition to reinvested earnings) from the U.S. to Latin American paper and allied products affiliates was $14 million in 1966, primarily for paper and paperboard mills. Net capital flow to lumber and wood products affiliates was $4 million in the same year.

Contribution to Income and Employment

Total local expenditures of the U.S.-controlled affiliates in the paper and allied products sector in Latin America were some $182 million in 1966. Table 10 shows the major breakdown of these expenditures. Foreign taxes took $13 million and direct compensation of local employees added about $33 million. In addition, $128 million were spent locally on materials and supplies.[8] In the case of lumber and other wood products (also shown in Table 10) total local costs were some $11 million, with local payroll expenditures of around $3 million. Local taxes were about $1 million.[9]

The total payroll of some $36 million supported almost 16,000 persons, with about 80 percent of the employees being accounted for

[8] According to the U.S. Department of Commerce study, "Paper Industry Leads," Table 20, parent companies of affiliates in the paper and allied products sector shipped some $20 million worth of goods "for further processing" to their Latin American affiliates. This accounts for the paper sent from the U.S. for conversion in Latin American converting plants. It also includes some shipments of pulp to foreign affiliates.

[9] Again, it should be emphasized that these figures only include those majority-owned U.S. affiliates which employ more than 100 persons, so that some of the activity in lumber and wood-based panels is not included.

Table 11.   Number of Local and U.S. Employees in U.S. Majority-Owned, Forest-Based Affiliates in Latin America, 1966

Employment category	Paper and allied products	Lumber and wood products	Total
Local			
Managerial	303	18	321
Technical and professional	499	46	545
Other	4,085	194	4,279
Wage earners	9,096	1,586	10,682
TOTAL	13,983	1,844	15,827
U.S.			
Managerial	25	7	32
Other	43	8	51
TOTAL	68	15	83

Source: U.S. Department of Commerce figures from May 1970.
Notes: The table includes data only for firms with more than 50 percent U.S. capital and with employment of more than 100 persons. The table therefore understates total employment effect.

by the paper and allied products affiliates (mean compensation was $2,250 per man per year). U.S. employees—mainly managerial and technical—accounted for only about one-half of 1 percent of the total (Table 11). Much of the technical work, including important decisions relative to project development, is in the hands of local personnel. In several cases, the only full-time foreign employee was the project manager, although in most instances the back-up administrators were trained in the United States, and temporary foreign personnel were brought in. In other cases, even the managers were from the host countries (HCs), with one or two U.S. specialists in other key positions.[10]

The HCs are interested in the productivity of the foreign capital and the new activity which it generates locally, including particularly the extent to which new domestic employment opportunities are created. The latter depends on the capital-labor ratios involved, the linkage effects, and the extent to which positions are filled by foreign or local persons.

In terms of the types of forest-based activities studied, the capital-labor ratios vary drastically, from an extremely high but variable one in the case of pulp and paper projects (between $150,000 and $300,000 per employee) to a very low one ($500 or less) for some types of

---

[10] However, in some cases, U.S. employees do not show up on the affiliate payroll and are not counted.

Table 12. **Labor Requirements for Felling and Crosscutting of Logs in Natural Tropical Forests**

Method	Normal daily production (m³)	Crew size
Axe	3–7	1 worker
Crosscut saw	11–17	2 workers
Power saw	25–60	1 operator and 1 helper

Source: Government of Great Britain, Overseas Development Administration, *Project Data Handbook*, section 7, p. 28.

forestry operations. There is some flexibility in the factor proportions which can be efficiently adopted in most forest-based activities. However, often logistical and efficiency considerations require introduction of some capital-intensive technology, with a corresponding reduction in direct employment opportunities per unit output.

The question of appropriate technology for forestry operations in the tropical high forest is by no means a clear one. Even the most modern companies tend to use fairly labor-intensive silvicultural methods, and most of the companies do not have their own felling and extraction crews, preferring to buy wood from independent contractors who have tended to be very labor intensive, using mainly axes and crosscut saws. As indicated in Table 12, labor productivity increases markedly with introduction of power saws—from 3–7 m³ per man-day with an axe to some 12–30 m³ per man-day with a power saw. If we assume a cost of $300 for a power saw with a three-year life span, and a 10 percent rate of interest, the depreciation charge per day (based on 250 operating days) amounts to about $0.46 (with no salvage value). Adding on fuel, spare parts, maintenance, etc., associated with the operation of the power saw, brings costs to around $1.00 per day. Assuming a conservative daily pay of about $1.00 for a man using an axe, a skilled man using a power saw could still be paid up to three times as much, or $3.00 per day, before the cost per cubic meter was equal to that for axe felling.[11] Naturally, the higher the relative wages, the more advantage there is to using power saws and other mechanical equipment to increase labor productivity.

---

[11] This follows from the fact that productivity using the power saw is about four times that for the axe. Adding on the additional $1.00 per day for depreciation, fuel, etc., brings us to $3.00 for the wage portion. Table 12 also indicates typical labor requirements for various silvicultural activities associated with utilization and management of natural tropical high forests in the Amazon, which is probably as difficult an area to work with as any in Latin America. (See section entitled "Contribution to Development of Technical Forestry.")

A major contrast between Latin America and Asia in terms of U.S. foreign investment in the forest-based sector is the fact that, in Latin America, most of the output of U.S. affiliates is sold in the region, that is, the regional market is a major factor in the investments. In Asia, most U.S. investment in the sector is associated with export, either to Japan or to the United States, generally via processing facilities in third countries in the case of log exports. One of the implications of high local sales of primary and secondary products is higher local multipliers and forward linkage effects. At the same time, imports by the paper and allied products affiliates from parent companies amounted to less than 10 percent of total costs of production in 1966.[12] Again, backward linkages from paper to pulp and to primary raw material subsectors are important in relative terms. These linkages are associated with further employment and income and stimulation of some local activity to supply goods and services to the foreign affiliates.

Estimates of local income multipliers for the forest-based sector are lacking for Latin American countries. However, estimated multipliers for the timber-producing sector in various regions of the United States are generally quite high. For example, one of the more detailed recent studies calculated a local income multiplier of 3.45 for the timber-producing sector of a rural county in northern Minnesota.[13] Other studies indicate multipliers in the 2.5 to 3.5 range.

One study dealing with income generation for forestry in Trinidad found that for each dollar's worth of standing timber (measured in terms of royalty values) another $7.31 was added by harvesting activities in the case of sawtimber.[14] This increase is based on royalty payments for timber and is therefore comparable in concept to the figures developed for the United States. However, with low royalties, added income generated is obviously much higher. In terms of employment, some 2,626 man-days of labor were required per 1,000 m³ of timber produced.

Foreign Exchange Contribution

The Latin American countries have a forest products import bill of around $400 million which they want to reduce. Many of the countries

---

[12] Local expenditures amounted to about 70 percent of total sales value in the forest-based sector in 1966.

[13] J. Hughes, "Forestry in Itasca County's Economy: An Input-Output Analysis."

[14] M. Gane, "Forest Harvesting in Trinidad."

**Table 13. Derivation of Net Foreign Exchange Effect of U.S. Forest-Based
Affiliates in Latin America, 1966**

($1 million U.S.)

	Paper	Wood products	Total
Exports generated by foreign investment (X)	+20	+3	+23
Imports generated by foreign investment (M)[a]	−44	−2	−46
Imports substituted by foreign investment (Ms)[b]	+245	+11	+256
Gross investment from foreign sources minus dividends, depreciation, interests, etc. transferred abroad (Ig − D)	+14	+4	+18
Net positive foreign exchange effect[c] (E)	235	16	251

Sources: May, *The Effects of the United States;* U.S. Department of Commerce, *U.S. Direct Investment Abroad—1966, Parts I and II;* and Lambiase, *U.S. Direct Investment Abroad.*

[a] It is assumed that M equals 15 percent of total sales (see May, *The Effects of the United States,* pp. 13 and 15).

[b] Ninety percent of total local sales of affiliates are assumed to be import substitution.

[c] Based on formulation in Chapter 3: $E = X + Ms + Ig − D − M$.

are short on foreign exchange, and their exports tend to be concentrated within a very few sectors and products. They have a strong desire and incentive to increase and diversify exports and to promote import substitution.

Of the total sales of U.S. Latin American affiliates in paper and allied products in 1966 ($292 million), only 5 percent went to the parent company or other affiliates. Further, most of the sales of the paper and allied products affiliates were within the country in which the affiliate is located, with only $3 million going to other Latin American countries and $4 million to the European Common Market.

The net foreign exchange impact of U.S. affiliates in Latin America in the forest-based sector is equal to export revenue (X) plus import substitution effects (Ms) plus net capital inflow (Ig − D) minus foreign exchange production costs or imports generated by the projects (M). [Net capital inflow equals gross investment from foreign sources (Ig) in the year being analyzed minus earnings remittances and capital repatriation (D).[15]] As an approximation, the net positive foreign exchange effect due to foreign investment projects in the forest-based sector was thus some $251 million in 1966. (See Table 13 for derivation of this figure.) The ratio of foreign exchange earnings or savings to total sales value is about 82 percent.

There are indirect beneficial impacts of foreign investment on export market access for domestic firms. Over time, species and products

[15] This is the formulation discussed in Chapter 3: $E = X + Ms + Ig − D − M$.

from the Latin American countries have become known on the international market through efforts of U.S. companies. In some cases, new domestic firms moving into the sector have taken advantage of the foreign investor's (FI's) experience in getting a particular species or product accepted in the U.S., European, or other markets. One of the critical barriers to expansion in tropical hardwood exports from Latin America has been and still is the lack of accepted markets for many of the species found in abundance in the natural tropical forests of the region.[16] In the cases where the FI has paved the way for new species or products, the HC has benefited. For example, this is the case with virola (*Virola* sp.), a popular veneer and plywood species sold in the United States and Europe. It is now sold overseas by a number of domestic firms in Brazil and elsewhere.

Contribution to Development of Technical Forestry

A final question of interest is the extent to which FIs have advanced technical forest management in Latin America. There are three distinct management situations for the three types of forests with which FIs are associated in Latin America; they are (1) natural coniferous forests, (2) forest plantations, and (3) natural tropical hardwood forests.

Natural Coniferous Forests

In the case of natural coniferous forests, the management questions relate to (1) protection against fire, insects, disease, and man, (2) soil protection, (3) regeneration of stands after harvest, and (4) cultural treatments for stand improvement (for example, thinning, pruning, etc.).

Natural coniferous forests are very limited in Latin America, as in the rest of the developing world. We studied two projects in the Central American countries, but other than these—and the major one was recently abandoned—there are no significant U.S. investments in natural conifer forest-based projects in Latin America, and U.S.

---

[16] Several of the most experienced individuals involved in tropical hardwood development have quite clearly suggested that they believe the only way to develop viable industries in such areas as Amazonia is to involve firms that have captive markets for the output or, at least, assured market outlets and advanced marketing facilities.

investors have contributed little, if anything, directly to management of this type of forest.

Forest Plantations

In the case of forest plantations, whether they be coniferous or hardwood, the techniques and economics of management have also been well developed in Latin America. The main problems relate to (1) site selection, (2) species selection, (3) preparation of site and establishment of stand, (4) cultural treatment of stand, (5) protection, and (6) harvest.

All the U.S. companies in southern Brazil are involved in technical management and development of plantations. As such, they contribute to the nation's forest capital and technology. However, the level of plantation management in domestic firms in most countries such as Brazil, Chile, Argentina, etc. is such that the FI currently is not able to bring with him any substantially improved technology. Rather, the FI as well as the domestic company develops new innovations, new species with superior growth or qualities, etc. over time. In many cases, innovations have been introduced by domestic professionals and technicians working for the FI, that is, foreign capital provides the means for the local staff to exercise and demonstrate its abilities. For example, in earlier years, foreign companies in Brazil, such as Rigesa, Champion, Olinkraft, and others, introduced substantial improvements in forest plantations, including the planting on a large scale of the U.S. southern pines (mainly loblolly and slash). Much of this work was done with the direct cooperation of Brazilian foresters.

But this is only one aspect of the contribution of these foreign firms. For example, Champion do Brasil pioneered in introducing extension programs for local woodland owners, providing training for Brazilian forestry students, providing initial financial and other support for local, independent loggers, and giving away seedlings to local landowners. Over the first ten years of the last program, some 40 million seedlings were given away. (Now Champion charges a nominal fee for them.) This program affected some 220 farms with a total area of some 24,000 ha. They support the seedling program with technical services. While the self-interest involved is evident, this program has provided definite benefits to the local economy and an example which has been followed elsewhere.

Another interesting contribution, which indirectly owes its origin to U.S. influence, is a highly successful cooperative research and development program for plantation forestry in southern Brazil.[17] This program is sponsored by most of the major forest products firms in the region, including at least three foreign companies, and is funded jointly by the companies.

The U.S. subsidiaries in southern Brazil are characterized by fairly long association in Brazil. They are in the country to stay and have integrated into the regional economies. Their forestry activities have been developed in response to their own needs for cheap wood. At the same time, they are involved in a truly cooperative effort with some of the most advanced domestic firms, such as Klabin, which has an international reputation for being in the forefront of pulp and paper developments.

One final example of innovation by foreign capital is the Jari project in the Brazilian Amazon. This project, developed by the multimillionaire, D. K. Ludwig, will include one of the largest artificial forests in the world. Some 17 million trees had been planted by 1972, including the fast-growing Asian species, *Gmelina arborea,* and pines. The aim is for 100 million trees by 1980. In 1971 alone, some 25,000 acres were planted. The yield of *Gmelina* is about 30 m³ per hectare per year. Initially, this project ran into a number of technical problems, but now, after adapting operations to local conditions, the project is moving ahead rapidly. There is still no production on a commercial basis, but the technical developments which have been and are being introduced by the company should benefit Brazil and other countries with tropical hardwood forest lands. In this case, the cheap land and tropical environmental conditions are the real resource of value, with the natural forest cover providing the initial nutrient base, but no commercial returns from harvest.

Natural Tropical Hardwood Forests

The major unsolved technical problems exist in the natural tropical hardwood forests. There is little agreement on the appropriate approach to management of these forests, although they account for by far the major portion of the forest area in Latin America.

In the Amazon, for example, there are those who argue for clearcutting and replacement with exotics, including pines. Others argue in favor of replacement with native species. Another group argues for

---

[17] Instituto de Pesquisas e Estudios Florestais (IPEF). Known foreign company participation includes Champion do Brasil, Rigesa Celulose, Papel e Emabalagems, and Olinkraft Celulose e Papel.

enrichment planting and other cultural activities aimed at perpetuating the natural forest, with selection cutting or group shelterwood logging being the harvest method.

A major problem affecting management of natural tropical hardwood forest relates to the nature of the composition of most of the tropical forest areas found in the Latin American region. While some fairly high volumes per hectare of virola (*Virola* sp.), cativo (*Proria copaifera*), and other species exist in some areas, the greater part of the forest area contains a heterogeneous mixture of species, most of which have no current commercial value.[18] As a consequence, procurement operations are on an extensive basis, with little economic justification for developments such as roads and other more or less permanent infrastructure. Without roads, forest management becomes difficult from an economic point of view, even if the technical operations are well understood.

Several companies told us that the incidence of transport costs in total delivered log costs runs as high as 50 percent or more. Foreign companies in Latin America have introduced various innovations to overcome this problem. For example, Boise Cascade's operation in Colombia developed a barge with complete veneer-peeling facilities. Since logs are in general a high volume or weight-to-value product, transformation to a higher value per unit volume or weight nearer the source (the forest) can permit economic removal of greater quantities of logs that would, without such transformation, have been submarginal. The floating veneer mill processes lower-valued logs into core stock veneer which is then shipped by sea to Barranquilla where the main mill is located. Company officials indicated that they were planning to possibly introduce a second such processing unit. This type of innovation could provide benefits for tropical forest exploitation far beyond those realized by the company introducing the idea.

The HCs themselves sometimes create barriers to FI innovation in forest management of tropical hardwood forests. For example, in the case of the Philippines, FIs in the past were required to establish forest management services for the lands on which they were operating. However, because of requirements with regard to hiring local persons and because of the higher salaries which the FIs could pay, the government found that the companies were hiring away the best

---

[18] In some areas only 1 percent or so of the sawlog volume is economically exploitable, given transport costs, demand, etc. In general, Southeast Asia has the most homogenous tropical forests. Africa follows and Latin America comes last. (See B. Lamb, "Tropical Hardwood Forest Resources"; H. Knowles, "Investment and Business Opportunities in Forest Industrial Development of the Brazilian Amazon.")

men from the critically understaffed professional ranks of the forest service. As a consequence, and for other reasons too, the Philippine government has decided to undertake the forest management activities itself, charging the companies a fee in lieu of company forestry activities. This same problem—the hiring of the best technical men by foreign companies—is not uncommon in the Latin American countries.

The payback on forest management operations, whether domestic or foreign, mostly occurs far in the future from the time when the management expenditures and activities are undertaken. With the high discount rates associated with projects in the tropical less-developed countries (LDCs), and with the high degree of uncertainty surrounding investment in many countries (which is implicitly, if not explicitly tied into the acceptable rate of return), forest management activities do not in practice have very high priority with some LDC governments, although politically many of them expound basic conservation principles. If they can extract a higher rent or royalty from exploitation without management expenditures, and put that money to use elsewhere, they are likely to do so. In perspective, this is a rational approach in many cases.

The economic rationale is indicated in the following example. A few years ago the government of British Honduras planned substantial investments in reforestation in response to a drastic liquidation of its natural forest capital. However, a senior economic advisor of Her Majesty's Treasury of Great Britain suggested the following:

I appreciate the strong temptation to invest in something for which it is quite certain that the country is suitable, and how this is reinforced by the knowledge that present difficulties are so acute because of the heavy disinvestment in natural forest capital in the past. I accept further that the ultimate return on forestry is very large. But there is a strong prima facie presumption against a poor country investing a big proportion of its capital in a project which yields a return only after so long a period. When a peasant is hungry enough he eats even his seed corn. I do not suggest that British Honduras is in so desperate a position as that. Nevertheless, I do not believe that it can afford to wait anything from 40 to 100 years to get a return on its money.

The Conservator of Forests was good enough to provide me with the information necessary to calculate a very rough rate of return on capital investment in forestry. The results I obtained were 8 percent on the relatively quickly yielding pine plantations and a rather higher rate on bringing pine under protection for natural regeneration. It may be suggested that, since money can be borrowed at 6 percent, the investment is justified. But this is an irrelevant comparison. For it is not possible for British Honduras to borrow unlimited amounts at 6 percent; nor indeed at any rate of interest, however high. Since forestry is competing for capital with other urgent needs—many of which can offer a much quicker return—the 6 percent rate at which a limited amount can be borrowed is not relevant; the effective rate of interest which capital should earn is far higher; and by this standard forestry does not pay. The conclusion is all the more true for investment in hardwoods, where the period of the investment is much longer and the true financial return correspondingly lower.[19]

Commercial forestry developments need to be considered as an integral part of processing projects. Taken in isolation, forest management tends to be associated with rates of return below alternative rates, whereas, when considered in combination with processing, acceptable rates of return (in terms of Downie's arguments) may result.[20]

One such example of successful development of forest management in natural tropical forests which considers the overall economic point of view and which relates wood supply to final product is PULPAPEL. A subsidiary of the Container Corporation of America, PULPAPEL has developed a technique for successfully utilizing and managing tropical hardwood forests in Colombia on a renewable basis. They developed a type of road system for hauling out logs that has successfully overcome the critical transport problem. Characteristically, the heavy rains wash away roads or make them impassible and uneconomic to maintain, but PULPAPEL tackled the problem and developed an answer which works in their area (upland tropical forests). They cable logs to ridge tops and take the wood out on roads similar to the old corduroy roads. Further, the natural regeneration which occurs after harvesting meets company requirements. Under current conditions they utilize most of the species on an area (that is, they clear-cut, leaving a few seed trees). This is one of the few such operations in the tropics that has overcome its transportation problems successfully. Colombians consider it a credit to the company, although the technology has not yet been adopted elsewhere.[21]

---

[19] J. Downie, *An Economic Policy for British Honduras*, p. 8.

[20] Cf. Yale University, School of Forestry, "Financial Management of Large Forest Ownerships," particularly papers by W. L. Moise and J. A. Segur.

[21] This is also one of the only companies in the world that is pulping a large number of tropical hardwoods in mixture. The accumulated experience could be of value in the future.

chapter five **Interaction Between Host Country and Foreign Investor**

In Chapter Three we discussed the factors which motivate the host country (HC) and foreign investor (FI) to become involved with foreign investment projects in the forest-based sector. We also considered briefly the actual foreign investment experience in the sector in Chapter Four. Now, in the present chapter, we will attempt to bring these two considerations together in an analysis of the interactions between HC and FI which determine the extent to which their objectives are met. Specifically, after reviewing some basic concepts, we will look at the actual interaction process in Latin America and the forest utilization contract, which is by far the most common type of formal agreement between the FI and HC. Finally, with this as background, we will consider why new investments are scarce in Latin America, why there has been a relatively large amount of expansion of existing projects, and why some past projects have failed in the sector. All three of these considerations relate directly to the more fundamental question regarding the nature and outcome of the interaction between HC and FI.

Interactions in the Foreign Investment Process

The outcome of interactions between HC and FI depends on relative bargaining strengths of the investor and the HC. In turn, relative bargaining strength depends on the relationships between the minimum requirements and motivations of HC and FI, as well as on the advantages (resources, services, and incentives) which each can offer in terms of the other's objectives.

In Chapter Three, we discussed the factors which motivate both the FI and HC to become involved with foreign investment projects in the forest-based sector. Two points are worth noting about these motivations and their relationships. First, some of the motivations of one

47

party may be complementary with those of the other party, while others are nearly always competitive or conflicting. Second, since there are tradeoffs between the different factors or objectives motivating the HC and FI, it is the sum total of the interactions between these factors which determines the relative bargaining position of each of the participants and the final result of negotiations. It becomes a complex matter to analyze empirically, partly because objectives and advantages are not specifically spelled out in most cases and partly because the potential tradeoffs between any two sets of advantages are difficult to predict quantitatively.[1] Yet some systematic view of these relationships is critical to the development of a logical framework for analyzing the investment process and its outcome.

Major Types of Interaction Situations

We can start with the assumption that in any given situation, certain *minimum requirements* for both HC and FI must be met for productive interaction to take place. The country may have certain laws or requirements which have to be met if an application for a project is to be considered in the first place, while most corporate investors tend to have certain guidelines for minimum profitability requirements, ownership shares, etc., which, if absent, cause the corporation to go elsewhere with its project.

The various factors determining minimum requirements are shown in Figure 1.[2] On the one hand, changes in internal conditions in the companies (A)—for example, in management objectives, profit position, etc.; in home country conditions (B)—for example, markets, laws, interest rates, etc.; and in the rest of the world (C)—for example, other investment opportunities determine changes in company policies and minimum acceptable conditions.

On the other hand, changes in internal HC policy conditions (D)— for example, changes in government, political motivations, influence of past experience, etc.; in exogenous conditions in the country (E)— for example, resource availability (uncommitted), internal markets, factor costs, transport, etc.; and in the rest of the world (C) determine change in HC policies and minimum acceptable conditions for accepting foreign investment.

[1] See R. F. Mikesell, *Foreign Investment in the Petroleum and Mineral Industries: Case Studies of Investor-Host Country Relations,* p. 29; and Y. Aharoni, *The Foreign Investment Decision Process.*

[2] A detailed discussion of the minimum requirements themselves can be found in Chapter Three.

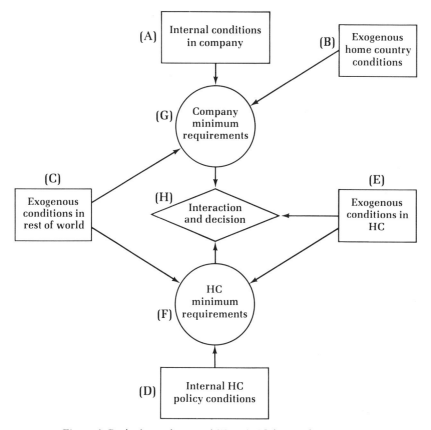

Figure 1. Basic determinants of FI and HC interaction outcome.

Finally, changes in HC and FI minimum requirements (*F* and *G*) interact with changes in the exogenous conditions in the HC, thus determining the type of interaction situation (*H*) which arises in any given case.

There are two possible situations in which minimum conditions of either or both HC and FI may not be met in any given instance. First, the particular physical, technical, or economic characteristics of a project may be such that one or more of the requirements cannot be met. Second, a minimum requirement of one party may be in direct conflict with, and prevent achievement of, the other's minimum requirement. For example, the HC may require by law that majority ownership of a project be in local hands, while the FI also requires majority ownership. Both situations may result in what we call a *direct conflict* situation. If one or both of the parties do not relax their minimum requirements, then the project is abandoned.[3] If one or both do change in a favorable direction—so both parties can meet their minimum requirements—then interaction can continue.

This leads to a second type of situation where the minimum requirements of both parties are met and some leeway between the two exists. We call this a *negotiable* situation. The final settlement in this case depends on negotiation or bargaining. For example, the HC may have a requirement that the *maximum* allowable repatriation of net earnings is 15 percent of registered capital per year. (Any lesser percentage is permissible.) On the other hand, company policy requires a *minimum* of 12 percent.

Finally, a third type of situation exists where both parties have complementary objectives, and minimum requirements of both parties merely reinforce each other. This is called the *complementary* situation. For example, the HC might want and require *at least* 50 percent, and preferably more, of production to be exported, while the FI is really interested in exporting the total output.

Summary of the Interaction Process

The interaction process is briefly summarized in Figure 2, which presents schematically the basic elements involved in the HC-FI interaction. We start with identification of basic HC and FI objectives (*A* and *B*). While it is possible to conceptually relate the two sets of objectives, it is more meaningful to identify and relate the factors which motivate FI and HC in terms of their objectives (*C* and *D*). These factors can be more easily identified empirically, and must be identified in order to analyze the interactions between HC and FI, that is, negotiations between the HC and FI revolve around the factors which motivate them to pursue foreign investment and not directly their objectives. This is related to the fact that there are tradeoffs between the conditions which can result in the same level of attainment of the objective. For example, for the FI, profits relate to total unit cost relative to price. It can achieve the same level of profit with higher direct cost of inputs combined with greater incentives (higher prices, fiscal incentives, etc.) as with lower input costs and less incentives. Along the same lines, there are tradeoffs between export receipts brought into the country by the FI and remittance of earnings.

---

[3] In the case of an existing project, it could result in expropriation-nationalization.

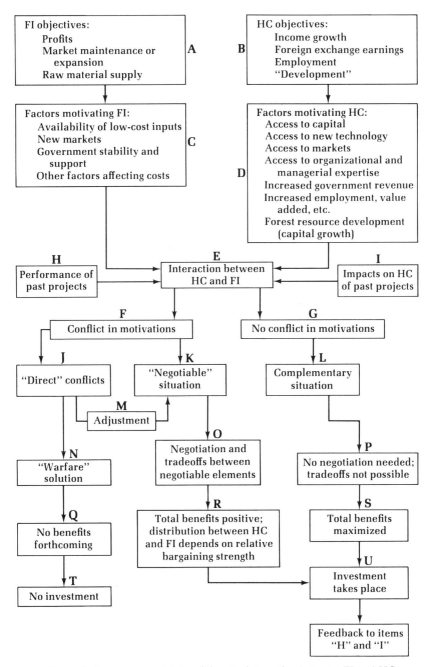

Figure 2. Summary model describing the interaction between FI and HC for a given investment proposal.

While we are interested in the internal tradeoffs for the HC or FI, the main interaction of interest is that which exists between factors motivating HC and FI (*E*). Here, we have identified for each set of motivating factors two possibilities which are analytically distinct at each stage of interaction. On the one hand, there might be conflicts between the factors (*F*), and on the other, there might be none (*G*). The latter category presents no problems, and there is no reason to consider complementary motivations further here—they do not provide bargaining advantages for either HC or FI (unless the complementarity is not perceived by either or both parties, and in this case they are not classified as complementary in our framework).

If a conflict exists between motivations, then we have the two possibilities mentioned earlier—the direct conflict (*J*) and the negotiable (*K*) situations. If a direct conflict is defined by inflexible laws or company policies, it results in a termination of the project and is of interest to us only in terms of how past direct conflicts can help us to avoid them in the future.

Whether a negotiable situation exists initially or is created, this is the type of interaction which is of greatest interest. It is in this category that we find the tradeoffs which ultimately determine the outcome of most negotiations.[4] And the outcome of the negotiation process determines *ceteris paribus* the relative impact and performance of the foreign investment.

Figure 2 represents a static situation or one point in time. Over time, as the interaction for any given project evolves, the exact specification of the minimum acceptable levels of achievement of objectives may change with changing conditions. The result may be a change in the type of interaction situation which arises. A negotiable situation may degenerate into a direct conflict, or vice versa; a complementary situation may develop into a negotiable one. Basically, any of the factors shown in Figure 1—exogenous conditions in the rest of the world, internal HC conditions, internal FI conditions, and conditions in the home country of the FI—may all change with time to influence the magnitude of minimum requirements in any given case.

More specifically, four main phases in the investment process can be identified (Fig. 3). Changes in interactions may arise (in response to the minimum requirements) from changes in conditions and from one phase to another. For example, Mikesell discusses the differences

---

[4] As mentioned, there are a number of factors exogenous to our framework which enter into the determination of the final outcome, etc.—the duplicity and perspicacity of the bargaining parties and the extent of information available (that is, the difference between the perceived and actual situations), etc.

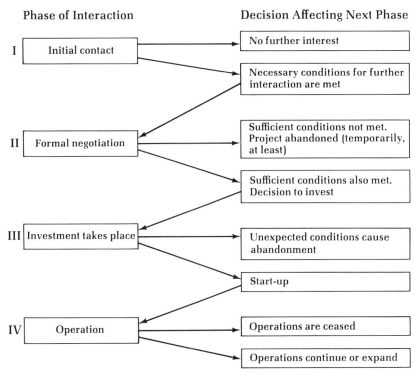

Figure 3. Phases in the interaction between HC and FI.

between the supply price of capital including risk *before* an investment in a new area is made and a commercial reserve of minerals or petroleum is proved and the supply price of capital including risk *after* initial development of a reserve for which adequate information exists.[5] During the initial contact and negotiation stages, when the nature and extent of a reserve is imperfectly known, the HC and FI might agree (through negotiation) on a mutually acceptable, high rate of return to the FI that allows for the risks involved. In the fourth, or operational phase of the investment process, the country might revise upward its view of the share of total rent or benefits which it should receive. This results in a new negotiable situation, or possibly a direct conflict situation which could result in expropriation-nationalization.

Very much the same type of situation exists in the case of forest-based investments, although often for somewhat different reasons.

[5] Mikesell, *Foreign Investment,* p. 35.

Initially, negotiations regarding development in a virgin forest area may result in a favorable result for the FI, in the form of low stumpage fees, tax concessions, etc., all of which have the effect of lowering costs for the FI and raising profit prospects to reflect higher uncertainty. The bargaining strength of the FI might lie in a combination of technological advantages in terms of exploitation, greater willingness to assume risks, market access for tropical hardwoods, etc. In addition, the country might recognize initially, at least, the low opportunity cost involved in exploitation of such forest resources.

However, as the project proceeds, as more becomes known about the resource, as local persons are trained, and as more information on the techniques for profitable exploitation becomes available, the country might revise its view with regard to the division of rent from forest exploitation.[6] This revised view can be manifested in a number of ways—through increased taxes, through substantially higher stumpage fees (either at the time of utilization contract renewal or imposed unilaterally on an existing contract), and through increased requirements for reinvestment in building up the renewable forest capital being exploited.

How could this latter approach add to the country's share of the rent? One answer is related to the fact that, with a few notable exceptions such as Brazil, most foreign investment projects in the forest-based sector of LDCs operate on utilization contracts for government lands, that is, the company does not own the land.[7] If the forest capital is built up during the life of the contract, then the country is left with an increase in forest capital when the contract terminates. An improved forest resource provides the country with a stronger bargaining position to extract more rent when renegotiating the contract.

We have now considered conceptually the main elements of concern in analyzing the actual interaction between HC and FI. First, we looked at the determinants of minimum acceptable requirements for productive interaction. Then we described the three possible types of interaction—the direct conflict and the negotiable and the complementary situations. Finally, we looked at the phases in the investment

---

[6] On this point, cf. R. Vernon, *Sovereignty at Bay: The Multinational Spread of U.S. Enterprise,* pp. 26–27.

[7] Utilization, or concession contracts, are by far the most common form of formal agreement between FI and HC, specifying the distribution of rents from exploitation of forest resources, the conditions of sale and development (e.g., of processing capacity which will provide increased local value added), etc. Because of their importance, both as an indication of the outcome of FI-HC negotiations and an indication of HC emphasis on various objectives, etc., we will discuss them in more detail in the section entitled, "Forest Concessions and Management Contracts."

process and considered how minimum requirements and, consequently, the type of interaction may change over time. Before proceeding to an analysis of the actual interaction process for the forest-based sector in Latin America, we need to mention that an additional factor in determining the outcome of interactions between HC and FI is the nature of the mechanisms available for negotiating and administering a contract once a negotiable situation is in effect. Here we are specifically referring to the institutions, bargaining skills, types of legal contracts, etc. which the HC has available and uses in its interactions with the FI. In Latin America, at least, these mechanisms are important in determining the types and numbers of projects brought to fruition.

The Interaction Process in Latin America

Each Latin American country has a somewhat different procedure which the HC and FI follow in negotiating an agreement to operate in the country, either through a subsidiary or some other form of affiliation, such as a joint venture. In most cases, with specific reference to developing an agreement for exploitation of public forest resources, treatment of nationals and foreigners is similar in law, although in practice the degree of scrutiny of a project proposal and the application of certain criteria from legislation related to foreign investment may make the treatment quite different.

One of the basic problems in the eyes of the investor is the great number of agencies involved from the time of initial contact to the actual operation of a project.[8] When we add on the red tape generated by the individual agencies, the total process of reaching a viable agreement becomes in most cases a confusing and time-consuming task. It generally requires the help of a local law firm and often the services of an "expediter," or middleman, who knows how to move requests through the maze of government bureaucracy. All of this adds to the initial costs of getting established. And this in turn discourages the investors.

In a number of countries the red tape and bureaucratic confusion does not stop once the forest-based project has been approved and becomes operational. In many cases, certain operating decisions— changes in production levels, changes in prices, etc.—need government approval, and such requests must pass the same route of agencies and individuals. Oftentimes, the result is that the company avoids

---

[8] This problem also exists in countries outside Latin America such as Indonesia. Cf. Government of Indonesia, *Technical Guide to Capital Investment in Forestry.*

changes except for those of absolute necessity. If the company does seek the HC's permission for a given change, then it generally costs a great deal in time and money.

Schmithusen lists some of the institutions that may be connected with preparing and granting forest concession or utilization agreements. They are as follows:

i.   The Head of State, the Cabinet or the National Assembly, if the approval of one or more of these is required for the establishment of a long-term contract.

ii.   The Minister or Ministers responsible for forestry and forest utilization.

iii.   Institutions concerned with the overall planning for economic development such as a National Development Council.

iv.   The ministries of Economics and Finance and/or national investment boards, if special investment incentives are to be granted, and Ministries of Transport, Communications, Power, Industry, Water Supply, Lands, Labour, Education, in relation to infrastructural requirements or development.

v.   Special boards such as foreign capital investment boards.

vi.   The Provincial or State Governments, in the jurisdiction of which the area to be granted is situated.

vii.   Institutions representing the interests of the local population such as Native Authority Councils.

viii.   The state forest service, and/or the technical agency that is in charge of forest contract matters.

ix.   The sections of the forest service which are responsible for the preparation of the agreement.

x.   Representatives of established forest industry.

xi.   A special working commission concerned with the selection of a possible grantee.[9]

As an example of the process and priorities involved, an FI goes through the following steps in Colombia before a final agreement is signed:

1. A general proposal is presented to INDERENA (the agency which manages the public forest resources). This general proposal states the nature of the proposed activity (project) and requests permission to proceed with more detailed planning and studies.

2. Once the general proposal is approved, the FI receives a permit (option) to inventory and prepare a management plan proposal for the forest area in question.

---

[9] F. Schmithusen, *Handbook on Forest Utilization Contracts on Public Lands,* p. 83.

3. This management plan is again presented to INDERENA for technical certification.

4. If certified, the proposal goes to government lawyers who check the legal implications.

5. It then goes to the Office of Planning (Planeacion) and to the Institute for Foreign Trade (INCOMEX) for review.

6. If they approve it, then the final decision is made and the agreement is signed.

The whole process can take up to eighteen months and involves a lot of shuffling between agencies and individuals. However, there is a great deal of flexibility in the final form of the agreement, depending on the planning office's priority for the project and the evaluation of INDERENA concerning its technical desirability.

Order of priority of a project is determined on the basis of explicitly stated criteria. The factors considered are:

1. The net effect of the project on the balance of payments.

2. The contribution of the project to employment.

3. The technological complexity of the project and the degree of initial and later utilization of domestic raw materials and parts or elements already manufactured or intended to be manufactured in the country.

4. The relation of domestic capital to the project and the proportion between the capital imported and the need for permanent investment and working capital.

5. The improvement of competitive conditions of the national markets.

6. The contributions to Latin American integration.[10]

The conditions considered most important are the balance-of-payments contribution and the level of employment. Consequently, in the evaluation more weight is given to these two factors. In addition, preference is given to projects which result in a diversification of exports. Forest management is a requirement for all projects.

Most other Latin American countries have similar procedures except for Brazil which does not give long-term concessions on public forest lands. The Brazilian philosophy is that the investor should own his land (and buy wood from independent dealers). Owning land, they argue, gives the investor an incentive to manage the forest resource on a long-term basis and tends to create an attitude of permanence in the project.

[10] Consejo Nacional de Política Económica. Resolution 9, 1968.

In many Latin American countries, the rigidity of the government and the complexity of the interaction is due to the fear that the investor will take advantage of the country in some way. This perception of the situation is understandable, given the past experience of many countries and the lack of information on the actual operations of foreign companies.[11] However, at some point, if the country wants to encourage foreign investment, it needs to increase its flexibility, clarify its policies, increase its monitoring system, and consider the policies and constraints on the side of the potential investor. (This point is equally true for the company wanting to invest.)

The country's various technical shortcomings and its lack of experienced government negotiators shows up in its policies and in its bargaining with the FI in establishing a viable formal agreement. Recognition of the country's desire to have a given project is generally expressed in the set of incentives and conditions which it offers the investor so that he in turn will agree to certain points which the country considers critical. The incentives commonly found in contracts in Latin America are many and often complex, with some of them being general and others being specific in the sense that they are tied to specific conditions of performance on the part of the FI. The types of incentives commonly used include reduction in or elimination of taxes, import or export duties, special exchange control regulations, and guarantees and subsidies of various sorts.[12]

Forest Concession and Management Contracts

The most important form of agreement between an FI in the forest-based sector and a Latin American HC is the forest concession agreement, or utilization contract. This is similar in many respects to mineral and oil exploitation agreements. However, in other respects they differ quite sharply because of the different nature of the resource being exploited.

Most forest utilization contracts contain provisions related to the following items:

1. Principals to the agreement and its duration.

---

[11] In the case of Honduras, recent evidence indicates that exporters (primarily foreign) were drastically underreporting both quantities shipped and prices received. The result was a loss to the country and a drastic change in policy and law (see p. 82). This is an extreme case, but is perhaps an indication of "things to come" in other countries.

[12] Cf. H. Gregersen, "Export Development Programs for Forestry."

2. Timber and other values conveyed.
3. Location of area involved.
4. Conditions of contract (for FI).
5. Measurement and payments.
6. Obligations of government.
7. Execution and recording of contract.[13]

The FI tends to be particularly concerned about the provisions related to measurement, payments, and obligations of the government, while the HC is more concerned with the conditions of the contract relating to FI obligations over time.

A complete and detailed discussion of forest utilization agreements and how they are developed in LDCs is available elsewhere,[14] and there is no need to repeat it here. Conflicts between the HC and FI can arise over most of the provisions contained in such agreements, but the major items which have caused disputes in the past relate to the determination of areas to be included, forest management practices, installation and timing of processing facilities, forest fees (including those for specific purposes, royalties or stumpage fees, and one-time-only area fees), and questions related to government responsibilities in terms of management and provision of infrastructure. Generally, most of the items of interest fall into the negotiable category, although such items as length of contract, installation of management systems, etc. are often specified by law.

Conditions of Contracts in Latin American Countries

The conditions and provisions of forest utilization contracts presently in existence in Latin American countries vary widely in terms and conditions and in benefits granted the concessionaire. For example, the length of contracts vary from short-term cutting permits (of, say, one year or less) to a type of contract that is essentially given in perpetuity if the concessionaire meets the requirements provided in the agreement. This latter contract provides the greatest flexibility, but it also creates some uncertainties for the company involved, since it

---

[13] Cf. S. Somberg, *Timber Sales Contracts for Latin America,* pp. 21–29. Schmithusen (*Handbook,* pp. 99–122) gives a detailed listing of clauses and provisions for long-term forest utilization contracts. He includes detailed examples of Mexico, Nigeria, and the Philippines. See also J. J. McGregor ("Forestry Concessions in the British Commonwealth Countries") who relates terms of concessions in British Commonwealth countries.

[14] Schmithusen, *Handbook.*

Table 14. Duration and Possibility of Renewal of Long-term Contracts as Determined in the Legislation of Various Countries

Country	Maximum duration (years)	Renewal possibility	Special remarks
Latin America			
Ecuador	30	No	
French Guiana	18	Yes	
Mexico	29	Yes	
Venezuela	50	No	Contract granted for 40 years
Colombia	20+	Yes	
Africa			
Congo-Brazzaville	25	Yes	Former "Permis temporaires d'exploitation" granted up to 25 years. Present "Permis industriels" granted for 10 years.
Cameroon	5	No	Contract to be renewed several times. Special renewal conditions for companies with wood conversion units
Ghana	Up to 50	No	
Ivory Coast	15	No	
Madagascar	20	Yes	Renewal only for one period
Nigeria	25	No	
Asia			
Khmer Republic	Up to 20	No	
Indonesia	20–30	Yes	Renewal proposed but not specified
Philippines	25	Yes	Renewal only for one period
Thailand	15	Yes	Renewal to 30-year period proposed
Pacific Area			
(Eastern New Guinea)		No	
British Solomon Islands	20	No	
Western Samoa	20	Yes	Renewal limited to 40 years

Source: Schmithusen, *Handbook*, p. 46. Colombian information from personal interviews.

is less binding on the government. (However, if the government wants to cancel or alter the terms of any type of contract, there are always means available.) Table 14 provides information on contract duration for a number of LDCs.

Most of the contracts provide for payment of fixed fees during their lifespan although a few include provisions for automatic adjustments in timber fees or for renegotiation at periodic intervals. Most also provide for renewal privileges, but in practice such a provision is often not taken too seriously, since there are always means for getting around this provision, if some more favorable opportunity arises.

Another common provision calls for consideration of the rights of local populations living in the concession areas. This is necessary from a practical political point of view. However, the extent to which the provisions are enforced depends more on local pressures than on the details provided in the contracts. In general, the main problem involved with most of the provisions in the existing contracts is their lack of enforceability due to inadequate forest service staffs to supervise them in the field. Many of the contracts are made for substantial areas (50,000 ha. or more), and the areas involved are generally inaccessible and poorly defined on the ground. Indeed, one concern of investors is the general lack of clear boundaries on the concession areas and the problems which arise with regard to illegal cutting and encroachment by local loggers and squatters who move into virgin forest areas to "high-grade" them (that is, take out the best logs) and/or clear the forest for the purpose of developing agriculture.[15] Shifting agriculture requires extensive land. As the productivity of one area is depleted, the shifting cultivator (or "squatter") moves on to another forest area, repeating the process of forest destruction.[16]

Despite the legal documents which provide clear definitions of the areas to be included in a concession, most governments and companies are helpless in keeping out squatters. As a consequence, many concessions are let but never developed, since the concessionaire finds new groups of squatters on his area daily. These people tend to follow the construction of new roads, and it is not uncommon to find them following directly behind the bulldozers as new roads are extended into the forest (as happened during the construction of the Rama road in Nicaragua). Although some governments attempt to force squatters off the land, their efforts are seldom effective.[17]

One U.S. company takes the view that direct bargaining with the local squatters is the best approach. It is argued that if the company can provide local populations with certain benefits, based on the exploitation of the forest resource, then the local inhabitants will not only respect the forest development project, but will also participate in its protection. The company involved has received criticism from the government, which favors more forceful action to eliminate the

[15] Naturally, this condition also means that it is easy for a concessionaire to cut from nonconcession lands. However, wood removed (and paid for) is generally measured at the mill, and this reduces the incentive to cut on nonconcession lands.

[16] FAO has estimated a loss of forest land to shifting agriculture of some 5–10 million ha. per year in Latin America.

[17] Honduras is trying a new approach, namely, to give local populations an economic incentive for protecting forest lands.

problem (partly because large private landholdings are also involved). The foreign company argues, however, that the government has not developed an effective way to deal with the squatter problem on a permanent basis. As soon as the army moves out of the area, the squatters move in again, and most governments cannot afford to permanently leave large contingencies of army personnel in remote forest areas.

Most of the current utilization contracts now include provisions for sustained yield management of the resource and for a schedule of developments which includes the eventual development of processing capacity. The latter provision often creates conflicts between the HC and the FI, since the government is generally interested in developing processing capacity as soon as possible, while the investor favors putting off such a commitment until the resource has proved itself.

Comparisons with Other Regions

The type of concession agreement typically found in a Latin American country is not unique to that part of the world. Even in the case of a developed country such as Canada, the provisions of concession agreements for utilization of public forests are very similar to those reviewed for Latin American countries. For example, an agreement between the Province of Alberta and Northwestern Pulp and Power, Ltd. provides for sustained yield management, a contract duration of twenty-one years, guaranteed renewal if conditions are met, and requirements for construction of processing capacity, etc. Most of the other provisions are also similar to those found in Latin American contracts, including types of fees to be paid, schedule of fees over concession period, government obligations for protection, company obligations to local persons living in concession area, etc.[18] Although the details differ, many other concession agreements in Canada and elsewhere (for example, the African and Southeast Asian countries) have the same general types of provisions (although the periods covered, specific fees, etc., may differ).

In nearly all Latin American countries, the concession system started with simple exploitation agreements that provided a company with a reasonable assurance for a continuous raw material source and provided the government with a pure rent in the case of natural forests that would not otherwise have been exploited. As development pro-

[18] Information on Canada was obtained from Mr. T. Clarke, Director, Forest Economics Research Institute, Canadian Forestry Service.

ceeded, countries became more concerned about assuring future supplies of wood; they became more aware that forests are "renewable"; and they became aware of the developed world's popularized term, "conservation." All these considerations have led to concession agreements with an emphasis on management and regeneration of the forests, whether such requirements are economical or not. The terms of agreements are still generally fixed over a long period; and royalties to the government are low, or at least that is what the governments come to think, as nationalistic interests expound on the theme of developed capitalistic countries "taking advantage of" the poor countries. Therefore, governments push for contracts with higher timber fees, land rents, etc. However, even in Canada, stumpage rates as low as $0.20 per cubic meter have been paid for timber in areas with marginal access. These fees are comparable to those paid in many LDCs. The point is that concession agreements need to include realistic fees for the timber cut which reflect the extraction and processing costs involved and the price of the final product. Stumpage fees should generally be related to the residual left after all other costs have been subtracted from the selling price. This residual may be very low.

To the investor who is going to install processing capacity, pay for a road network, etc., the question of stability and guarantee of tenure and provisions for economic renewal of a concession agreement is more important than the exact fee. In most countries today, the sizes of individual concessions are large, and although this requirement (by most corporate investors) is not generally emphasized, it is another factor of key importance to the investor.[19]

Negotiating Agreements—A Summary

A concession or utilization agreement between an HC government and an FI in the forest-based sector forms the basic legal document governing a project. It contains restrictions on the FI and guarantees negotiated benefits to him in terms of incentives, guarantees from the government, etc. In some cases, there may be additional agreements signed which relate specifically to fiscal incentives, foreign exchange requirements and restrictions, etc. However, such additional agreements tend to be standard for foreign investment in all sectors, in the countries in which they are applied.

The specific nature of the restriction and costs imposed on the FI and the benefits, including incentives, granted in concession agree-

[19] On these points, cf. McGregor, "Forestry Concessions."

ments are generally determined (within certain limits imposed by law or government regulation) through bargaining between the government and the FI. The limits tend to be narrow when it comes to technical management/exploitation specifications and are somewhat broader with regard to the final determination of incentives offered and specific costs imposed for wood and land rents. The determination of the final set of costs and the magnitude of the incentives depends on the bargaining strength of the parties involved and the tradeoffs they are willing to make. Often there is only one potential investor interested in a particular forest concession possibility, while in other cases there may be several investors competing with each other. From the government's point of view, the latter is generally preferable, since it puts the government in a better bargaining position (at least in theory). However, governments sometimes have the false notion that they have a monopoly on forest resources, and this is hardly ever the case when it comes to tropical hardwoods, nor is it so when the project will be relying mainly on plantation-grown wood. While several companies might be interested in a potential project, most of them also have other alternatives. In several Latin American countries, governments have lost opportunities for development because they have acted on the assumption that the potential investor has no other alternatives. In point of fact, it tends historically to be more the opposite—the government finds itself with only one firm alternative for a particular project or area. Not infrequently, a project area has been defined for a long time and despite tentative inquiries on the part of various investors, none have actually applied for a concession.[20]

To explore the bargaining for concessions and the resulting agreements more fully, we need to look at (1) the reasons behind the lack of new investment in the sector in recent years despite a great deal of looking and "nibbling," and (2) the failures of past forest-based projects in Latin America that have occurred because direct conflict situations arose. We will start with the former point which is closely related to the latter.

Why New Investments Are Scarce

In the past five years, there has been a lot of interest in new opportunities within various Latin American countries on the part of

---

[20] Much of the problem also stems from the fact that governments often do not know what their resources are worth. This problem will be more fully discussed in Chapter Six.

U.S. corporations, but almost none of this interest has borne fruit (for example, projects considered by International Paper in Honduras, Litton in Guyana, and others in Surinam, Nicaragua, and Ecuador have been unsuccessful).

How do we explain this dearth of current new investment activity? Given the assumptions that both investors and countries have some knowledge about existing opportunities, and that both have devoted some efforts to initial exploratory activity, the implication, in the context of the framework developed earlier, is that more direct conflict interactions have arisen in recent years. As will be recalled, this type of interaction is defined as one where the HC and/or FI find that their minimum acceptable requirements for one or more aspects of the potential investment project cannot be met, and, furthermore, neither party is willing and/or able to adjust its minimum requirements to the extent that a negotiated agreement could be reached.

What changes in the countries and companies have caused the increased number of direct conflict situations? There are two possibilities which need to be explored. First, is it a matter of country and company policies which have become more restrictive? Second, is it a matter of changing external conditions—for example, changes in markets or resource opportunities—which have narrowed the range of total benefits to be shared to the extent that the total is not great enough to meet both HC and FI minimum requirements, even if more restrictive policies were not adopted by the parties involved?[21]

In past years, many forest industry investments were made in Latin America by even the large U.S. corporations with little regard for overall corporate "strategy." More generally, there was a lack of such overall strategies to guide domestic as well as foreign investments.[22] Based on our interviews, it appears that most companies within the past decade have become more sophisticated in terms of planning their investments on the basis of an overall corporate strategy.[23] At the same time, corporate officers have also begun to recognize and to accept the fact that they cannot enter most countries without serious consideration of the HC's objectives and its claim to a fair share of the rent or royalties obtained from forest exploitation.

While the companies still have the same general objectives and motivations, the above factors make consideration of new investment

[21] These were causes mentioned in Chapter Two. It could, of course, be a combination of the two. Changing opportunities in other parts of the world could be a contributing factor to changing policies.

[22] See Chapter Three for further discussion of this point.

[23] Most of the corporate officers interviewed emphasized the importance of this factor in their current deliberations over foreign investment projects.

proposals a much more complicated task involving detailed feasibility studies and severe scrutiny. While, as mentioned earlier, most companies do not necessarily compare many alternatives before making their investment decisions, they apparently do scrutinize individual proposals much more closely and conservatively than they did in the past. To the extent that substantial risks and uncertainties exist for a given new proposal—and this is usual, rather than the exception—a company is much more likely to reject a project unless it promises extremely high returns. And given recent trends in Latin American demands (see below), acceptable returns under any given set of conditions are less likely to be forthcoming.[24]

Changes in U.S. corporate strategies and objectives have partly been a reaction to the recent trends in HCs toward more restrictive demands on the FIs.[25] One of the more widespread changes in LDCs is the increased emphasis on local processing of raw materials. There are few countries now that will accept an FI who proposes setting up a log export project with no plans for installation of processing capacity. Many countries have outright restrictions against log exports.[26] The inclusion of processing capacity in a project requires a commitment of much greater amounts of fixed (risk) capital in the HC. If the potential investor has the normal fears about expropriation and other uncertainties, he is much less likely to commit himself and thereby reduce his flexibility.

Despite complaints from the developed countries, LDCs have legitimate reasons for restricting log exports in many cases. If the product is being exported or is an import substitute, processing at home creates more local value added and provides more employment and greater foreign exchange earnings or savings. Much depends on the conditions in the HC, the willingness of investors to commit capital to establishment of local processing facilities, the economic availability of prerequisite factors of production, infrastructure, etc. However, if the country's requirement for local processing capacity is unacceptable to the FI, and if, because of this, the resource is not utilized, then both parties may lose.

---

[24] We have already mentioned a number of other considerations related to the FI's home country restrictions, e.g., generally unfavorable economic conditions at home, etc.

[25] This topic has been the subject of discussion for a number of years. Cf. United Nations, FAO, Pulp and Paper Advisory Committee, *Some Obstacles to a Private Investor in Establishing Export-Oriented Forest Industries in the Developing Countries,* and FAO Committee on Wood-Based Panels, *Obstacles Impeding the Flow of Investment Capital to Forest Industries in the Developing Countries.*

[26] For example, Paraguay, Guatemala, Chile, and others.

An increasing willingness is evidenced on the part of U.S. corporations to undertake processing in the HC, particularly in labor-intensive operations such as veneer slicing. Yet, in many cases, the old motivation for investment—to supply home processing facilities with cheap raw materials—is still important and conflicts with the HC objective. A compromise, which has worked to the advantage of both parties in many cases, is to do the primary processing in the HC (for example, to produce core stock veneer or molding blanks) and then carry out the final stages of processing (for example, plywood) in the FI's home country (as is done by Georgia Pacific in Brazil and Champion International in Peru).

In the case of natural tropical hardwood forests, the FI has some legitimate arguments for wanting to start with log exports and then gradually work into processing (if conditions eventually warrant it). Out of the hundreds of species found in a given tropical hardwood forest in Latin America, only ten or so may be commercially accepted, and even fewer are accepted on the export market. This means that very low volumes per unit area can be harvested and that unit costs are correspondingly high. An objective in this case is to get more species known and accepted. But companies are generally not willing to undertake the risk or the task of testing and introducing new species unless they have the flexibility to take them out in log form and process them in already-existing facilities.[27] Some countries, however, refuse to let companies export logs, even for test purposes; and many of the countries that are emphasizing domestic processing will not allow companies to export logs on a commercial basis before installing processing facilities. Officials say that the following occurs: The company comes in under an agreement allowing it to export logs for, say, five years, after which it is required to establish primary or secondary processing, followed five years later by final processing capacity. However, after five years, when the company has high-graded the forest, it announces to the government that it overestimated the resource, which will not support economically the installation of processing capacity. At least one major, otherwise favorable, project was lost in Central America because of misunderstanding or disagreement over the *potential* for this type of situation to arise. In another case, a concession agreement was terminated when the processing capacity was not installed on schedule.

While Latin American forest resources seem virtually limitless from

<hr />

[27] On the problems involved in introducing new species, see J. L. Stearns, "The Problems Encountered in Introducing a New or Unknown Wood on the American Market."

a physical point of view, they have been severely depleted from an economic point of view. Even in the vast forests of lower Amazonia, loggers have to travel far to find concentrations of virola. Many of the best, accessible concentrations of cativo, virola, etc., in Latin America have either been depleted or are already under contract. This is a prime example of an exogenous change in country conditions which affects the interaction situation directly, regardless of whether there are changes in HC or FI policies. When minimum acceptable conditions are fixed and the "pie" to be shared has diminished (that is, the availability of timber and profitability of exploiting it is reduced), then a direct conflict situation may arise unless the HC and/or FI reduce their minimum requirements regarding royalties, restrictions, taxes, etc.

A second change, which was mentioned earlier, is the increasing demand on the part of HCs for more control and larger shares of the profits from projects, with no comparable willingness to undertake the risks involved. Countries often demand the same restrictions on profits for a new, exploratory project with high risk as for an existing one which is being expanded. This problem has been treated elsewhere,[28] and we need not delve into it here.

Several companies suggested that the increased squeeze on profits was turning them into "safe" investors: they are willing to pay others to do the initial work on a forest concession and take the major risks, accepting in turn a lower rate of return themselves. (The question is, Are there enough "others" who are willing and competent to do the initial development work on a concession?)

One specific example of a country creating such a direct conflict is Honduras. In their 1972 forest law,[29] they require that all new projects utilizing public timber resources must include the government as major stockholder (more than 50 percent ownership), and further, the government can use the public forest resources for the project as part or all of its share of the capital. For private investors, this condition is unacceptable, since it would essentially amount to the minority stockholder putting up 100 percent of the risk capital, although having 49 percent or less of the control and the profits. Other less glaring instances of the same type of requirement exist; for example, in Andean Pact countries.

A third change concerns labor costs faced by the investor. Cheap labor has traditionally been considered one of the motivating factors that brings foreign investors to the Latin American countries. How-

---

[28] See Chapter Three, and Mikesell, *Foreign Investment,* p. 35.
[29] Article 96 of Decree 85.

ever, a group of experts on pulp and paper development suggest that there are few opportunities in pulp and paper in LDCs in which it is possible to realize any significant savings in overall labor costs as compared with projects in the developed world.[30]

While this view is not held by all, a problem mentioned by most companies is the increasingly high social cost of labor which has to be borne by the company. In one case studied, the actual wage paid is only half of the total bill for labor, the remainder being made up of various social costs for housing, insurance, etc. In another case, the FI complained of the "womb to tomb" philosophy of some governments which requires that once a person has been hired it is the responsibility of the company to "carry him to death." This type of policy creates a lack of flexibility for the company and often discourages a prospective investor.[31]

We can now sum up the discussion with regard to why more new investments have not taken place (or why the rate of new investment in the forest-based sector of Latin American countries has been lower in recent years). First, there are fewer large concentrations remaining of currently economically accessible species (for example, virola, mahogany, cativo), and most companies, because of their typically high overhead costs, are only willing to get involved in fairly large projects with guaranteed wood supplies over time.

Second, internal conditions in the major investing companies have changed. There is much less investment on the basis of senior executive "hunches" and much more concern for quantitative analysis and systematic comparison of alternative opportunities based on quantitative data. Since reliable information on opportunities in the LDCs is often lacking, this creates a major stumbling block and a bias against projects for which little information is available.

Third, increasing demands on the part of LDCs for a larger share of the returns from exploitation of their natural resources, coupled with more demands on foreign investors to establish local processing facilities, create situations where projects which would previously have been acceptable are now marginal. With lower expected rates of return, risks must be lower; and they are not. (The solution here is not necessarily to reduce HC demands—which are quite legitimate in many cases—but rather for the FI to make adjustments and the HC to provide other encouragement to investment, such as is the case in Brazil.)

---

[30] United Nations, FAO Pulp and Paper Advisory Committee, *Some Obstacles,* p. 4.

[31] Particularly during the early stages of a project, management needs the flexibility to hire and fire in order to adjust the production process and obtain an acceptable group of workers.

Finally, as development and social concerns advance in Latin America, earlier advantages in terms of low labor costs, tax advantages, etc. disappear, and the investor has less incentive or motivation to get involved.

Changes in objectives, motivations, and exogenous conditions have been such in recent years as to create an increasing number of direct conflict situations, with the natural consequence being a lower rate of new investment in the sector. Chapter Six discusses some solutions to these conflicts, but in order to gain further insights on the problems or obstacles involved, we need to look at the reasons behind failure of U.S. forest-based projects in Latin America.

Failures of Past Projects

We looked in some detail at five failures, defining "failures" to include projects for which firm commitments were made and at least initial funds were spent in the HC (regardless of whether the project ever became operational), but which have since been abandoned. In addition, we looked at two operating projects which are presently in a situation where they will in all likelihood close down operations in the near future. We are interested in the extent to which the interaction between HC and FI resulted in failure and the nature and causes of the direct conflict(s) which arose to trigger the failures. Table 15 presents information on the main causes of failure as determined through interviews and examination of documentation on the projects.

Based on available information, it appears that there was no one

Table 15. Main Causes and Incidence of Failure Among Projects Studied

Main causes of failure	Incidence
Change in government policy or uncertainty with regard to future policies[a]	2
Inadequate information on resources	2
Technical problems which raised costs	2
Market problems	1
Change in company policy	1
Lack of adequate government support in providing infrastructure, technical services, etc., which were agreed upon	1
Avoid advice from locals	2
Lack of FI control or interest	2

Source: Based on information for seven projects obtained from interviews, and company and HC documentation. Countries involved include Chile, Ecuador, Colombia, Honduras, Nicaragua, and Brazil.

[a] Includes one case of expropriation.

problem that was common to all the cases. In two instances, there were clear changes in government policy which created the direct conflicts. In two other cases, lack of adequate initial information on the project's concession resources contributed to higher than expected costs and lower profits. This problem is apparently a common one encountered in other parts of the world,[32] and it has several policy implications which will be discussed later. In both cases, the resource involved was a natural tropical hardwood forest. In one instance, experienced persons warned the company about the resource problem, but the company did not heed their advice.

Problems stemming from a disregard of local advice occurred in several projects. For example, HC foresters warned one company about using heavy equipment on the thin tropical soil. The company disregarded the advice and cleared land with large heavy equipment, planted seedlings, and found that they did not grow because the nutrients had been removed in the land clearing. In another project, local professionals cautioned the foreign company about using bulldozers in swamp logging. The company disregarded the warning and lost expensive equipment. There are other similar incidents as well.

For one project, an important factor contributing to its abandonment was a marked change in company policy associated with internal changes in the corporate structure. The new decision makers viewed the uncertainties of the project in a different light than had earlier officers, deciding to take a more conservative stand with regard to commitment of further funds to the project, which had an estimated 12 percent rate of return.

Another company stated that its estimate of the future market was overoptimistic at the time of initial investment and subsequent conditions caused it to delay further commitment of funds as per the contract with the government. The company expected the government to accept the delay on the basis of its problem. However, the government took the opportunity to cancel the concession agreement. This particular situation was also tied up with a change in government and a revised policy with regard to the desirability of foreign participation in forest resource development. Essentially, certain high government officials wanted to develop the resource themselves through establishment of a public corporation.

Finally—and somewhat paradoxically in terms of HC complaints about excessive foreign control—for two projects the main cause of failure appears to have been a lack of adequate parent company in-

---

[32] Cf. DeBoer, "Impressions of Industrial Problems in Tropical Forestry."

volvement or interest in the projects.[33] In both cases, initial poor judgment led to problems which could have been avoided with more supervision and support of the projects by the parent investors. When the problems developed, the FIs preferred to abandon the projects. One of these projects is now being revived by another investor. (This last point brings up the related observation that in several projects that failed, other investors took over and revived them.)

In sum, it appears that most of the same factors are important in explaining failure or abandonment of projects and the lack of new investment. The main factors include extent of government stability and support, changes in HC conditions, and amount and quality of information available on resources, costs, market conditions, and profitability prospects (which are tied up with a number of factors such as comparative technical efficiency, cost levels, etc.). In addition, failures and abandoned projects are associated with internal changes in the investing corporations and lack of sufficient interest and control on the part of the parent company.

New Investment in Expansion of Existing Operations

We mentioned earlier that most of the new foreign investment in the forest-based sector in Latin America is concentrated in expansion of existing operations. Additional discussion of the factors responsible is important in order to understand the conditions associated with successful ventures (see also the discussion which follows in Chapter Six).

It seems clear from our interviews that an important factor influencing the decision to expand existing operations has been the lower level of uncertainty surrounding such investments as compared with investment in new countries and new projects.[34] The lower level of uncertainty is associated with several factors, all involving better information availability. First, and as a general observation, an existing operation continuously generates reliable information which is in line with the requirements of the FIs for making decisions on new projects. For example, characteristics and availability of the resource being utilized are better known from actual operating experience. Yields per hectare,

---

[33] This point is supported by P. Arnold, "Why Development of a Tropical Forests Industry Can Be Difficult." He also mentions the general problem of inadequate planning.

[34] It should be emphasized that it is more difficult to isolate reasons why something was done than it is to determine why something was not done or why something failed.

species composition and characteristics, etc. can be better estimated, as can costs of procurement. In some cases, the subsidiary has created its own wood supplies (plantations) and has clearly in mind expansion possibilities, perhaps already owning the land for expansion. While problems with local populations—including particularly the squatter problem—still exist, the limits of the problem and its possible solutions are also better known. Expansion investment can take advantage of all these factors in a way that investment in new projects cannot.

Second, experience has been gained regarding the appropriate technology for use in extraction and processing. Such experience is often the only way to actually come to grips with particular problems in a country. The costs of obtaining such experience in new countries and new projects can often be very high. The effects can be felt in terms of costs of production as well as investment, and problems of over-capitalization (which has created difficulties in the past) can more easily be avoided.

As mentioned earlier, most of the projects in Latin America are oriented to the local markets of the region. To the extent that previous experience can provide information on such markets, the FI is able to avoid certain initial costs and reduce the level of uncertainty surrounding a new marketing program in a new country where previous information on markets is typically nonexistent or inadequate.

While relations with a host government do not necessarily improve over time, the continuous presence of an FI generally has the effect of clarifying the "rules of the game" and the nature of the relationships involved. As will be discussed later, it is not necessarily the severity of policies that bothers an investor as much as uncertainty surrounding future changes in such policies which would adversely influence the results of long-term planning efforts. As such, the stability and greater degree of certainty with regard to expected relationships that tends to come from actual presence in a country is appealing to the investor. It gives him the opportunity to make a better estimate of what will be involved in his new investment.

Finally, most of the successful foreign investment operations in the forest-based sector of Latin America have become integrated into the local (national) economies of the countries in which they exist. (Expansion investment will obviously only take place in already successful ventures.) Local personnel are employed, contacts with local law firms, etc. are well established, thereby easing the way for new investment. As the old companies integrate into the local environment, they develop a lower profile and subsequently a reduced political uncertainty. Additional investment in the old subsidiary is seldom as noticed locally as a new investment project of the same magnitude would be, and the risk of friction developing is less.

These are some of the factors which lessen uncertainty surrounding expansion investment, apparently making expansion investment more attractive than new projects to investors.[35] The question remains, Are such investments comparable in other respects, particularly with regard to profitability? While we could not obtain enough quantitative information to let us answer this question in a systematic fashion, there is no indication in the information available that the profitability of expansion investment in the forest-based sector of Latin American countries is less than that for new investments in new countries. Among other things, existing operations are for the most part located in countries in which the full market potential and resource potential are considerably above existing activity. There is still room for profitable expansion.

Regardless of how such investments compare, the evidence indicates that, at the present time, companies are favoring expansion investment rather than projects in new countries. In Chapter Six we will discuss the implications in terms of policies for overcoming bottlenecks to an expanded contribution of foreign capital to development of the sector in the Latin American countries.

---

[35] See Chapter Three for a discussion of changing FI policies and their greater conservatism in recent years.

chapter six **Implications for the Future**

So far we have analyzed past and existing U.S. foreign investment activity in the forest-based sector of Latin America. We looked at the nature and magnitude of such activity, the factors which motivate host countries (HCs) and foreign investors (FIs), and the interactions, problems, and conflicts that have resulted in a slowing down of new investment activity and in failures of past projects. In this final chapter, we will discuss some implications for the future, concentrating on those policies and measures necessary to overcome obstacles to further development of the sector which appeared from our investigation to be most important and most likely to have an influence on foreign investment activity.

Opportunities for Forest-Based Development

There is a growing consumption of wood and wood products world-wide. To meet their expected future requirements, most developed countries will have to expand imports. A proportion of the new import requirements will likely come from Latin American countries. In the United States, imports of wood products in general are expected to double by the year 2000, from the substantial amount of 2.4 billion ft.[3] (round-wood equivalent) in 1970.[1] Projected U.S. imports of hardwood lumber by the year 1980 are some 400–600 million bd.-ft., an increase of 100–300 million bd.-ft. over the 1970 level of imports. Hardwood plywood imports (mainly tropical) are forecasted at some 3.3–4.1 billion ft.[2] in the same year versus 2 billion ft.[2] in 1970.[2] These

---

[1] Cf. U.S. Department of Agriculture, Forest Service, *Outlook for Timber in the United States.* It is expected that *net* dependence on imports in the United States will increase from 8 percent of consumption in 1970 to some 15 percent of projected demand by the year 2000 (under assumption of rising domestic prices).

[2] U.S. Department of Agriculture, Forest Service, *Outlook for Timber,* pp. 182 and 184. Some 60 percent of U.S. hardwood plywood consumption is presently imported.

are sizable amounts, and expectations are that they will increase further by the year 2000. While Asia has been the traditional supplier of tropical hardwood products to the United States, there is a limit on the extent to which that region can continue to expand exports to the United States and other areas.[3]

Growth in the Japanese market can be described along similar lines except nearly all the increase in requirements will be imported. We have already mentioned that Japan is moving into the development of the Brazilian forest sector in a major way, with the main aim being exports of eucalyptus-based pulp and paper products to Japan. Japan will also increasingly compete with the United States for Asian tropical hardwoods.

Europe is also increasing its imports of pulp products, coniferous sawnwood, and tropical hardwood products, mainly sawnwood and molding products, as well as some veneer and plywood. Significantly, in terms of the problems mentioned earlier with species heterogeneity in Latin American forests, the European market is increasingly accepting a wider variety of species, many not previously exported from Latin America.

In the case of internal markets in Latin America, increased demands on local production will derive from new opportunities for import substitution, the steady expansion of the Latin American market, and the existence of regional country groupings which provide advantages for intraregional trade and local production. Compared with 1965, the regional demand for sawnwood is expected to more than double by the year 1985. An even higher increase in the demand for wood panels has been forecasted, and the demand for pulp and paper is expected to double with each decade.[4]

All the above demands suggest opportunities for expanding both local and foreign investment in the forest-based sector of Latin America. An idea of the size of these investment opportunities in Latin America can be derived from a study carried out by Eklund.[5] He estimates that investment in new forest industry capacity and replacement of existing capacity to meet increased requirements will increase

---

[3] This is because of limits on available forest resources, a rapid conversion of forest land to agriculture, and increasing competition from other importing areas. On these points, cf. U.S. Department of Agriculture, ibid., pp. 137–138; Tuolumne Corporation, *Market Study for Forest Products from East Asia and the Pacific Region;* and K. Takeuchi, "Tropical Hardwood Trade in the Asia-Pacific Region," pp. 59–60.

[4] United Nations, ECLA/FAO/UNIDO, Papers from the Regional Consultation on the Development of the Forest and Pulp Paper Industries in Latin America.

[5] R. Eklund, "Use of Capital, Labour and Forest Resources in the World Forest Industries."

worldwide from $37.9 billion between 1970–1975 to some $59.3 billion between 1980–1985.[6] He further estimates that the share of this total in developing countries will increase from 11.8 percent in 1970–1975 to some 16.2 percent in 1980–1985. In terms of annual compound growth, investment in the developing world would increase by some 8.1 percent, compared with a worldwide growth rate of 4.6 percent. In total, 75–80 percent of the investment will likely be in the pulp and paper industry.

Even if one takes a more conservative estimate than Eklund of necessary investment increases in the forest-based sector of less-developed countries (LDCs) in the 1980–1985 period—say, $1.0–1.5 billion per year—with the assumption that at least 30 percent will be in Latin America, mainly in the pulp and paper industry, investment would be substantial, say, around $300–450 million per year on the average in order to meet requirements.

It seems clear that a part of the increased requirements regionally and in the rest of the world will be met by the Latin American countries. The role that the Latin American economies can play in supplying the needs of the developed regions depends on the extent to which they can compete with the traditional sources of imports into these regions. While current levels of exports of tropical hardwoods from Latin America are low, Latin America's share is increasing, as we pointed out in Chapter Two. However, while regions such as the Brazilian Amazon are feeling the pressure from the outside world, it still remains that competition exists in the world tropical hardwood market for meeting the traditional uses and wood types (for example, the soft red-brown woods such as virola, meranti, etc. used in tropical hardwood plywood). Expansion of exports from Latin America will require that the countries have better access to markets, proper quality controls, volume output, and economic efficiency.[7] It will also require special efforts to introduce new species into the traditional markets.

In relation to the satisfaction of the increased local demands, new possibilities for import substitution have been created by advances in

[6] These investment requirement estimates are based on demand estimates by the United Nations Food and Agriculture Organization and estimates of required capital replacement rates.

[7] The cost advantages in Latin American countries over the developed countries lie in low land costs, better physical growth rates for wood, and lower costs in establishing plantations. The disadvantages lie in higher required infrastructure investments, higher industrial plant establishment costs, and higher current freight costs to the importing centers. For relative costs, see also United Nations, Economic and Social Council, Economic Commission for Latin America, *Account of Proceedings and Recommendations of the Regional Consultation on the Development of the Forest and Pulp and Paper Industries in Latin America.*

wood technology and chemistry which create opportunities for commercial exploitation of more species from the natural forests. For example, progress has been made in the utilization of mixtures of short and medium-length fiber species in the production of pulp and paper, fiberboard, etc. Also, new knowledge is being acquired in relation to the commercial management of natural tropical forests (see Chapter Four). Also, we have already discussed local production possibilities created by an expanding area of plantations of fast-growing species.

Expansion of local utilization and exports will affect both natural forests and man-made forests. It is likely, however, that major emphasis will be on plantations of fast-growing pines and hardwoods, such as the eucalyptus. It is clear that it is not always the existing forest cover that is important in creating viable projects. Rather the low cost of land, climatic conditions, etc. are the factors attracting investors in many parts of Latin America. A healthy sign of the times is the fact that foresters and to some extent politicians have been able to shed the traditional myth that because a large area of natural forest exists it inherently must be valuable for commercial development.

Foreign participation in the expected growth of the forest-based sector will depend in part on the extent to which investors and Latin American countries overcome the bottlenecks and conflicts enumerated here and in previous chapters.

Overcoming Obstacles to Expanded Foreign Involvement

If one accepts that there are opportunities for an expanded contribution of the forest-based sector to Latin American development and that foreign capital, technological and managerial skills, and market contacts can contribute to realizing such development potentials, what conclusions can we reach regarding the obstacles to be overcome or reduced in order to realize the potential contribution of foreign investment? Also, what are the policy implications concerning the means for overcoming existing problems?

It was concluded in Chapter Five that evolving company policies as well as changes in country policies and conditions which attract investors have contributed to an increase in direct conflicts between the HC and FI and have created obstacles to new foreign investment in the forest-based sector.

The changes in country policies are twofold: increased demands on the part of the countries for more direct control over the development of their forest resources and a greater share of the benefits from development without a corresponding willingness, in some cases, to assume an increasing share of the risks involved with such investments. Second, increased emphasis on investment projects which develop the resource on a renewable basis and commit funds to establishing local processing capacity, which, in turn, has tended to increase risks for the FI and sometimes reduced potential profits below levels acceptable to the FI.

Depending upon the magnitude of such changes in country policies, direct conflicts could develop with no changes in minimum acceptable conditions of the FIs or in their policies. However, there have been corresponding changes in the companies and their attitudes which have affected their policies and minimum acceptable conditions for investment. These include (1) stricter scrutiny of investment decisions to make sure they fit into an overall corporate strategy, and (2) more conservative views of the risks and uncertainty involved in foreign investment projects with a correspondingly higher minimum acceptable return on investment for a given level of risk or uncertainty.[8]

The interaction of the HC and FI in response to these changes on balance has been characterized by more direct conflicts, less room and flexibility for negotiation, and fewer projects (see Chapter Five).

At the same time, abandonment of existing projects has continued to take place. The factors found responsible include:

1. Changes in government policies (and uncertainty with regard to future policies).
2. Inadequate information on available forest resources.
3. Unforeseen technical and market problems.
4. Changes in company policies and lack of adequate interest in the subsidiary.
5. Company disregard of local advice (see Table 15).

In the remainder of this chapter, our interest will be focused on the means for reducing the various obstacles created by the above factors and the means for providing a more productive environment for developing and sustaining viable projects that will meet FI objectives and contribute to HC economic development.

As noted in Chapter Five, we want to look at the means for transforming direct conflict situations into negotiable ones. Figure 4A illustrates the problem of direct conflicts. *OT* represents the total to be shared, whether it be net returns, ownership percentage, total employ-

---

[8] In some cases it is a matter of the FI becoming more aware of uncertainties, while in others it is a matter of risks or uncertainties actually increasing because of changes in market conditions, policies, etc.

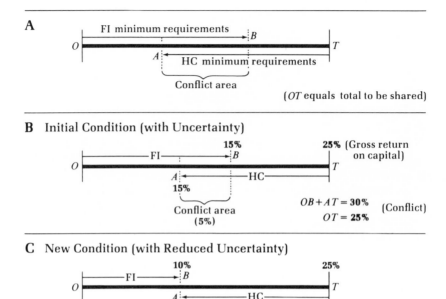

Figure 4. Interaction of HC and FI minimum requirements.

ment opportunities, etc. For example, in the case of ownership OT represents 100 percent, OB is the investor's minimum acceptable ownership, and AT is the HC's acceptable minimum. (In the case illustrated in Fig. 4, both require more than 50 percent, thus the conflict area AB).

There are two basic ways in which the conflict area (AB) can be eliminated. First, for some conflict items, the total area (OT) can be expanded while OB and AT are held constant. Second, minimum requirements $(OB$ and $AT)$ can be reduced (one or both of them). The first possibility is rather limited in scope, particularly in the short run. There is the possibility that if estimated net returns on a given investment can be increased, and *if* the FI accepts a fixed percentage on total investment, then, holding everything else constant, an unacceptable situation can be transformed into an acceptable or negotiable one. However, in the case of most conflict items—for example, percentage of HC and FI ownership, proportion of local employment versus foreign, etc.—there is, of course, no possibility of increasing the total to be shared (100 percent is the maximum possible). In any case, we have little basis on which to discuss this first possibility further.[9] Rather, we must concentrate on the measures to reduce the HC's and/or FI's minimum requirements (OB and AT), given the total (OT) to be shared.

The level of minimum requirements (as discussed in Chapter Five) is mainly a function of (1) alternatives available elsewhere, (2) noneconomic (political and psychological) factors influencing HC and FI motivations and objectives, and (3) perceived risk and uncertainty regarding a given project or investment opportunity. In the remainder of this discussion we will concentrate on the last item. While it would be interesting and worthwhile to look at the relative merits of the noneconomic motivations of FI and HC and to analyze alternatives available outside the region in the light of how these influence minimum requirements of both, this is beyond the scope of the present discussion.[10]

Looking at our more limited objective, we can argue that, for any given set of motivations and policies, the degree of uncertainty surrounding an investment is critical in determining the minimum requirements of the FI, and to some extent those of the HC, for any given project and, consequently, the extent of direct conflicts. Figure 4B and C provide an example of this. Given the basic policies of a certain company, the minimum acceptable level of return to the FI for a given project with a high degree of perceived uncertainty might be 15 percent (Fig. 4B) to take into account the perceived risks and uncertainty. Now if the level of uncertainty surrounding the project can be reduced, then the FI might reduce its minimum acceptable expected profit level to 10 percent (that is, reduce the premium for uncertainty). If the particular technical circumstances surrounding a project are such that an expected 10 percent rate could be obtained, after taking into account government requirements with regard to local tax payments, other contributions to domestic social welfare, etc. (AT), then there is the likelihood that a reduction in investor uncertainty, by reducing the company's acceptable profit level to 10 percent or less,

[9] Essentially, we assume here that OT is given and is the maximum possible for a given project at a given moment. As new information and new technology become available OT may be expanded and uncertainty reduced. But we can say nothing here about the nature of such changes in specific terms.
[10] In any case, there is little the HC can do to influence alternatives elsewhere, and political and psychological motivations (which determine some minimum requirements) are not related to any specific sector.

would transform a previously unacceptable or direct conflict situation into a negotiable one (Fig. 4C).[11]

Reducing Uncertainties

The uncertainties affecting minimum requirements in forest-based projects range all the way from purely political ones associated with changes in government policies to those that are almost entirely associated with technical aspects of a project (that is, politically neutral). While uncertainty is obviously not unique to projects in Latin America, there are some types of uncertainty which are particularly critical in many of the countries of that region. The ones which interest us particularly are discussed in detail in the sections that follow.

Uncertainties Related to Host Country and
Foreign Investor Policies

The countries of Latin America have been moving toward policies in which the HC retains a greater share of the potential benefits from foreign investment in the forest-based sector. At the same time they have been adopting policies which restrict the level of foreign control of projects. The effect of such changes has been quite different in different situations. In some countries, foreign companies have fully recognized the more restrictive nature of country policies but have invested anyway. For example, in Colombia which adopted the more restrictive policies of the Andean Pact, there has been an expansion of foreign (U.S.) investment in the forest-based sector. In the case of Honduras, the situation with regard to foreign investment was uncertain for several years. In 1972 a new forest law was passed (Decree No. 85 of February 10, 1972) which required majority ownership by the government in any new, large-scale venture involving use of national forest resources. Further, the law states that the government's share of the investment could be the forest resource. Passage of this law discouraged FIs. Then in January 1974, Law 103 was passed which established a new, semiautonomous government corporation which has control over management of all forest resources (public and private),

---

[11] The discussion in Chapter Five (pp. 72–74) indicates that one of the main reasons why new investment is concentrated in expansion of existing facilities and/or in countries in which investment already exists is because of the relatively lower level of uncertainty inherent in such investments.

development of forest industries, and all export marketing of forest products. The corporation (COHDEFOR) is well managed by highly dedicated professionals. Confidence in it and in the new forest policy is growing, as is the interest of FIs, and at least two definite proposals have been received for rather large projects.

It appears that, within certain limits, it is not necessarily the severity of policies which discourages the investor or raises his minimum requirements, but rather his assessment of uncertainty surrounding the future of such policies and his concern that fluctuations in policies will create conditions which make it impossible to effectively plan longer-term investment projects, such as those involving forestry.[12]

One possible way to reduce such uncertainty for the FI is for him to become involved in joint ventures, and even more so, in joint ventures where the FI has a minority interest along with certain contracted rights to manage specific critical aspects of a project. Uncertainty is reduced, since such ventures are usually considered domestic enterprises and have more local support.[13] Until recently, there has been little interest on the part of U.S. corporate investors in taking on minority positions in joint ventures in the forest-based sector despite the fact that many progressive companies in other sectors have had a fairly long history of relative success with them.[14]

There is one outstanding example of a joint venture in the forest-based sector in Latin America. ADELA Investment Company, a multinational corporation based in Brussels, which includes substantial U.S. participation, is involved with joint ventures. ADELA has recently invested in four new projects in the forest-based sector—a notable increase in terms of the overall distribution of ADELA activities (see Table 16). A significant point is that ADELA holds minority interests in all of these projects. Obviously, ADELA, which has considerable experience and local expertise in Latin America, not only supports the contention that there are opportunities in the forest-based

---

[12] Cf. E. Keller, "Opportunities and Problems of Private Investment," p. 4. This relates to what Keller calls the "cycles" problem in Latin America, or the lack of continuity in government policies, plans, and actions in some countries. He suggests that these cycles run between four and six years and are characterized by government changes accompanied by shifts in policies toward foreign investment. Lack of adequate plans and policies for periods of one to two years as the change in government takes place and the new policies are formulated, and increasing amounts of red tape due to the tight controls imposed as a new government comes into power. The basic characteristic is uncertainty and instability.

[13] The assumption is that there is more uncertainty surrounding majority-owned foreign enterprises.

[14] Cf. W. G. Friedmann and J. P. Beguin, *Joint International Business Ventures in Developing Countries.*

Table 16. Forest-Based Projects Invested in by ADELA Investment Company, as of September 30, 1972

Country	Type of project
Paraguay	Veneer and wood products
Mexico	Wood complex
Guatemala	Wood products company
Honduras	Pulp and paper (not operational)
Brazil	Rayon pulp; integrated pulp and paper mill; and sawmill and wood products company
Colombia	Manufacture of wood products (plywood, etc.)
Peru	Wood furniture manufacturing and wood extraction and processing
Venezuela	Furniture manufacturing

Source: ADELA Investment Company, S. A., *Monthly Bulletin*, October 1972.
Note: ADELA became involved in four of these projects since January 1969.

sector, but it also strongly feels that local development and control is desirable, including majority control by the non-ADELA partners.[15]

A related point of particular interest to the FI is that joint ventures with strong local participation can avoid common problems associated with foreign firms with regard to taxation, registration of patents, etc.[16]

Further, in some countries, notably those of the Andean Group, the existence of local activity in a sector or even local interest in investment is an important consideration in acceptance of FIs and treatment of wholly owned foreign affiliates after their establishment. Joint ventures can overcome this problem to some extent. As local Latin American interest in the forest-based sector increases, the joint venture approach may become the only realistic alternative for entering a promising local market or gaining access to a basic raw material source.

We identified a number of other uncertainties related to government policies affecting FI minimum requirements which could be reduced. Some involve varying interpretations of the concession agreements or long-term utilization contracts, which form the basis for most projects in the forest-based sector. In some cases, there is an initial unwritten understanding between the FI and HC concerning the interpretation

[15] Keller, "Opportunities and Problems." Strong technical support from ADELA is emphasized.
[16] One U.S. corporation official told us that the major problem they encounter in a foreign country is adjustment to the country's "unwritten laws and procedures." In some cases, it is expected that the company "support" certain officials and if they do not do so, it is taken as an offense. On the other hand, if they attempt to do certain other things in the "twilight" zone of the law, they are severely penalized. This is where local participation can be helpful and possibly essential.

of highly technical conditions which both parties recognize cannot be effectively met in practice due to the nature of the conditions existing in the HC (and the lack of adequate information on which to base actions to carry out contract conditions). After investment has been made, an initially loose interpretation by the HC of a given contract condition may be tightened up, possibly resulting in the creation of an economically impossible situation for the FI. Past experience with such situations is built into the expectations of the FI in setting minimum conditions.

On the other hand, the HC may have had similar bad experiences with an FI who has taken advantage of the lack of local HC experience with forest concessions and inadequately trained manpower to administer an agreement and enforce conditions. The lack of experienced negotiators in many countries has resulted in establishment of agreements which are unfavorable to the HC. Over time, as the HC comes to realize this, it attempts to make changes unilaterally to gain a more equitable situation. Several other examples of related policy problems encountered with concession agreements can be mentioned.

First, in some countries there is an unproductive use of concessions where they are obtained for speculation. While speculation is a legitimate fact of economic life, it is undesirable from the national viewpoint if, as a consequence, the concession stands idle for long periods of time as it changes hands. We found several cases where this was so. The solution to this problem—which has been adopted by many countries—is to stipulate conditions with regard to timing of activities related to development of the concession and conditions of transferral so as to avoid unproductive speculation.[17]

The question of speculation is closely allied to a second problem. Legitimate investors come to a country and are misled by promoters, often foreign promoters. In some cases, substantial sums of money have been lost,[18] and in others, it is merely a matter of a company losing confidence and leaving the country dissatisfied and without a project. The solution to this problem in each country differs. If government officials are involved in the misrepresentation, the problem is more difficult to solve. Although most large companies have technical and financial personnel and legitimate HC contacts who can overcome

[17] B. H. Payne and D. S. Nordwall, in *A Review of Certain Aspects of the Forestry Program and Organization of Indonesia,* also emphasize this point.

[18] For example, a major U.S. company invested in a concession development where the promoter was appointed their representative. Initial cost estimates were soon revised upwards. This occurred several times before the company abandoned the project.

such a problem, it is a point which is of great concern to smaller and medium-sized countries and investors who have no such facilities.

A third problem relates to overeagerness on the part of countries to have a forest area utilized. As a result, concessions are sometimes given to foreign companies notorious for the way in which they have neglected previous contracts. In one such instance, a government official stated that he knew the reputation of the company, "but it was the only company that bid on this area." While this argument might seem rational at the time—some revenue is better than none—there is an alternative: the area might be made more attractive to a legitimate firm given certain incentives.[19] If such an alternative is not followed, the ill will which might be created in the country could prevent successful negotiation later with legitimate firms in other areas.

Another aspect of concession or utilization contracts, which is important to the FI in setting his minimum requirements, is the length of the contract and the provisions made for renewal of such agreements. If it is expected that an investor will set up integrated extraction and processing facilities, there must be some long-term guarantees of wood supply. Appropriate clauses outlining conditions for renewability of contracts appear in a number of standard concession agreements used by various Latin American countries. Also, the exact nature of price- or fee-escalation clauses is important. If this question is not answered explicitly, uncertainty for the investor increases, since he reasons that when renegotiation time comes around he will have become committed, oftentimes with a sizeable investment that reduces his flexibility to negotiate favorable terms.

A fifth problem which has to be explicitly considered—but which is often treated in a superficial fashion—relates to the question of concession boundaries and the rights of local populations living within such boundaries. If it is expected that the concessionnaire will develop an effective forest management system on concession lands, it is important that he have the guarantees and flexibility to operate. The question of control of the concession area is particularly important to the FI if long-term management and utilization requirements are included in the agreement.

In many areas of Latin America, ownership rights for both private and public lands are not clear.[20] The result is, as we have observed,

---

[19] Such incentives naturally have to be considered as a cost to the country in calculating the potential contribution of the investment to national income and development. If the required incentive is too large, the project becomes uneconomical from the HC's point of view and should not be undertaken.

[20] The lack of preoccupation in establishing clear ownership rights on public lands is derived from the usually low value of these lands and the impossibility to control limits anyway, even if they are well established.

the existence of concessions granted under conditions of strong uncertainty regarding their real limits.

The solution clearly rests on an increased effort on the part of the government to produce clear and definitive specifications of ownership rights and concessionaire rights. Much effort is being put forth in this direction, starting with basic cadastrial surveys. However, in addition to government action, foreign companies need to study and define the conditions of ownership of the land with which they plan to be involved. This effort must necessarily be accompanied by exhaustive field work, since registries do not always describe the real situation.[21]

Even if land ownership is clearly defined, we still have in Latin America the problems derived from the practice of shifting cultivation, usually involving large numbers of peasants who move from one site to another as soon as the exploited land becomes eroded or loses its productivity and as new lands are opened up. In their migration peasants occupy public and private lands. Their eviction, which in many cases is politically undesirable anyway, is particularly difficult because they move fast and their number is usually large.

The solution to this problem obviously goes beyond the context of policies developed in relation to the forest-based sector alone. A reorganization of agriculture is implied, which in many Latin American countries means drastic changes in the whole structure of society and land ownership. If feasible, it is usually extremely costly in financial and political terms. As an alternative, we have seen that in some situations companies have found ways to identify the interests of the squatters with those of the investors, and thus, at least to some extent, they have prevented further losses of control over land on which they rely for their wood supplies. In those cases in which the investor operates on the basis of plantations financed totally or in part by him, the problem acquires critical dimensions. It is unrealistic for companies to develop areas where property rights are not well defined or where there exists an especially acute pressure for using land under the shifting cultivation system. Even when some problems can be reduced by the participation of local interest in the forest venture, the company cannot realistically expect to enforce its rights with the support of the HC's law enforcement agencies. This is also a particularly sensitive area for foreign company involvement. Therefore, the solution is either to stay away from these areas or to devise new and imaginative ways of dealing with the problems, including the possi-

[21] Honduras, recognizing the severity of the problem of land ownership uncertainty in developing its forest resources, recently passed a law that put *all* forest lands and their use under the control of a semiautonomous government corporation. Private individuals are to receive payments from the corporation for wood harvested on their lands (see pp. 82–83).

bility of explicitly recognizing the interests of the squatters who form a legitimate segment of the HC's population. The right of the company to deal directly with the squatters must be made explicit in the basic agreement with the country.

A final uncertainty which may influence the FI's minimum conditions for negotiation and agreement is the question of the timing and phasing of investments required under a utilization agreement. For example, to the extent that the FI is not permitted to bring out wood in log form in order to test it technically and on the market prior to committing sizable investment in processing capacity, uncertainty surrounding the investment increases and so do the minimum acceptable conditions. As we discussed this problem in earlier chapters, we pointed out that the HC faces a parallel uncertainty on the other side. If the FI is permitted to start with log exports, what type of guarantees will the HC have that the FI will not remove the best logs and then default on the agreement? There are, of course, means for providing safeguards so that this type of situation will not arise. However, negotiating such safeguards for the HC and balancing them off against safeguards for the investor is a complicated task requiring the skills of experienced and knowledgeable negotiators. And such skills are often lacking in the HC.

Some argue that the way to avoid the various problems mentioned above is to move to the short-term stumpage sale system such as that used in the United States. It is further argued that such a system will provide higher returns to the HC and will to a greater extent avoid the possibility of a country signing away benefits which rightly belong to it.

However, in the case of the national forests of the United States, the stumpage sale system works for a number of reasons not paralleled in most LDCs. Even if they could attract investors, LDCs generally cannot count on the conditions which would make this system work to their advantage. In contrast to the United States, they lack adequate numbers of technical personnel to lay out and administer sales, adequate information on which to base minimum bid prices, information on costs which are needed if the system is to work, access, and situations where they will get competitive bids, etc. And, in any case, the U.S. system has many problems which would only be magnified in most LDCs.

The concession approach, as currently used in Latin America, might not provide the government with the maximum theoretical short-term returns from its timber—for example, not as high returns as from specific stumpage sales to the highest bidder. But the government has to be interested in real and not theoretical returns, and this means attracting investment. Within the constraints imposed by the minimum requirements of interested investors for certain long-term guarantees, the government can and should attempt to negotiate to its maximum advantage. But, as mentioned, this will not necessarily result in maximum revenue, particularly if it means a lack of projects with long-term growth orientations.

Abandonment of the basic concession or long-term utilization/management concept would not appear to be in the best interest of most Latin American countries. Rather, recognizing the government's legitimate desire (or demand) to get a fair share of the returns from forest exploitation, the emphasis should be on improving information on exploitation and processing potentials and methods so that total returns and long-term development potentials can be increased and the HC and FI will mutually have better information, thus reducing the uncertainties perceived by both parties (and thereby some of their minimum requirements). In addition, the mechanisms for negotiating and administering contracts should also be improved.

Reducing Uncertainties Through Improved Flow of Information

Many of the policy-related uncertainties discussed in the previous section could be reduced by improved flow of information between the HC and FI with no changes in the actual policies and the ways in which they are established and imposed.

As discussed earlier, lack of appropriate information on forest resources, costs of production, and market potentials has been a contributing factor to failure of natural forest-based investment projects in Latin America.[22] Our investigation indicates that along with the

---

[22] The importance of the information factor in foreign forest-based projects has been emphasized by P. Arnold, "Why Development of a Tropical Forests Industry Can Be Difficult"; D. DeBoer, "Impressions of Industrial Problems in Tropical Forestry"; and indirectly by O. C. Herfindahl, *Natural Resource Information for Economic Development*, and by M. Nelson, *The Development of Tropical Lands: Policy Issues in Latin America*. See also Chapter Five. The problems of inadequate information are by no means an exclusive characteristic of the forest-based sector in Latin America. For example, N. Wollman, *Water Resources of Chile*, studying water resources projects in Chile, points out that "the main conclusion . . . will hardly startle anyone: 'efficient' or 'rational' choices regarding water use cannot be made without adequate facts and figures and Chile is short on both" [p. 202] . . . and "most of the unsuccessful public projects that have been built in the past owe their failure to the lack of reliable data . . ." [p. 206]. It is worth noting that Chile was chosen for this study because, among other reasons, "[it] has a long tradition of irrigated agriculture and with the possible exception of Mexico, it has better hydrologic data than can be found elsewhere in Latin America" [p. 1].

evolution of company policies it is apparently also a major factor contributing to the lack of new investment activity in the sector in Latin America. As the expected profit margin is reduced (due to increasing demands on the part of the countries, etc.) FIs require better information on which to base their investment decisions, since wrong information is more critical for a marginal investment than for one that provides expected profits considerably in excess of minimum requirements of the investors. (The acceptable margin for error is reduced.) Just as countries have become more sensitive to "exploitation" by foreign capital, they also have become more cautious and less willing to enter into contracts with foreign capitalists. We found reluctance on the part of both FIs and HCs to enter into productive negotiation in some cases because neither party felt it had adequate information on which to base its position in negotiations. This is encouraging in the sense that it indicates a greater awareness on the part of each party to the concerns and requirements of the other. However, given the trends in HC and FI policies, it will lead to less and less investment in the future, unless efforts are made to increase the flow of relevant information.

While it is obvious that more research and more and better information would be helpful, it is difficult for both FIs and HCs to evaluate their efforts in this area: There is no acceptable operational way to measure the relative benefits of various information expenditures except on an ex post basis in some cases.[23] Thus, the questions remain, What are the priorities for generating information? Who should initiate information programs, and who should pay for them?

Many millions of dollars have been spent generating information on the natural forests of Latin America. International organizations and governments attach high priority to information activities in the forest-based sector. For example, as of 1966, the United Nations Special Fund had provided support for a number of forest resource information projects in Latin America at a cost of some $11 million, or about 20 percent of its total contribution to activities related to information-generating programs in the region. In turn, the host governments of Latin America allocated an additional $14.6 million to these activities, or 22.3 percent of the total they dedicated to information projects (Table 17).[24] However, much of this effort has not yielded the economic information required by potential FIs and HCs alike. This is partly due to the fact that design and execution of such inventory and survey work has tended to be dominated by mensurationists and biologically oriented foresters rather than by those who might have

---

[23] Cf. Herfindahl, *Natural Resource Information.*
[24] Ibid., p. 84.

Table 17. Host Government and United Nations Expenditures on Investment
Information in Latin America, as of June 1966

Sector	Funds provided by		UN Special Fund (%)	Host govern- ment (%)
	UN Special Fund (millions of U.S. dollars)	Host govern- ment (millions of U.S. dollars)		
Forestry	11.0	14.6	18.6	22.3
Minerals	13.1	10.9	22.0	16.7
Fisheries	5.5	9.2	9.3	14.1
Soil	1.7	2.4	2.9	3.7
Meteorology and hydrology	8.1	7.7	13.7	11.8
Irrigation	5.8	9.4	9.8	14.4
Water resources	8.0	6.4	13.5	9.8
Other	5.9	4.7	10.0	7.2
TOTAL	59.1	65.3	100.0	100.0

Source: Herfindahl, Natural Resource Information, p. 84.

put information programs in the proper perspective to the requirements for economic development. Such an orientation is needed in
future resource information programs.[25] The information question
needs to be phrased in the following context: The objective is to
develop productive programs of forest utilization and development;
therefore, what information is needed to start activities that will move
toward this objective? Basic information on the forest, its composition,
volumes available, etc. is still needed, but the amounts and accuracy
of the information generated needs to be tied to the requirements of
potential projects and investors. Traditional mensurational information
needs to be supplemented by information on economic aspects of development (economic accessibility, operability conditions, harvest costs,
etc.) which in the past has not been adequately considered until after
negotiations have broken down or until after an established project has
run into problems.

One of the reasons for overemphasis of some information and under-
emphasis (or information voids) in other areas is the lack of a problem
or project orientation. Government information programs have been
set up with available manpower and their specialties in mind rather

[25] On this point, cf. United Nations, Food and Agriculture Organization, Report
of the Headquarters Meeting of Forest Inventory Exports on UNDP/SF Projects.
Partly due to similar criticism, recent UNDP/FAO projects in Latin America
have increasingly integrated resource information activities with forest management and development activities.

than in consideration of the specific development problems of the countries themselves. While we can conceptually derive sophisticated problem-oriented models for determining benefits and costs of expenditures in one area of information generation versus another, in practice, such models are difficult to apply. Yet, at least some general policy guidelines can be suggested for setting up adequate information programs and for avoiding misallocation of scarce funds to unprofitable activities.[26]

As mentioned previously, the government should focus on generating more information at the project level in order to stimulate an expansion of foreign and/or local investment in forestry development. Information programs might be phased so that by using existing information, it is possible to determine likely prospects for development areas within the next few years or within the medium-term planning period. Once the prospects have been identified, information generation should produce progressively more detailed information "until the decision maker would be satisfied that the margin of error had been reduced to the point where he could either proceed with development investment or abandon the project."[27] The judgment as to when the point of decision has been reached would obviously depend on the type of project, the groups involved, etc. The point is that information would be collected with specific developments in mind and not for its own sake.[28]

Second, there is a great deal of data currently available in the Latin American countries that is not translated into usable "information," that is, it is not collated and communicated in a usable form to the appropriate potential users, including FIs.

Much of the information already available in Latin America on the forest-based sector could at slight additional cost be channeled to local as well as foreign investors. Preston and others have suggested the establishment of an international or at least regional information pro-

---

[26] Cf. Nelson, *The Development of Tropical Lands,* p. 185. Nelson suggests that governments should fall back on cost-effectiveness as a guide in developing programs to reduce the degree of uncertainty surrounding investment opportunities.

[27] Ibid., p. 186.

[28] As Nelson (ibid.) suggests, "The accumulation of detailed information about natural resources that have no immediate prospect of development constitutes a serious misuse of scarce investment funds" [pp. 184–185]. See also Herfindahl, *Natural Resource Information,* pp. 193–194. Supporting this idea, Heinsdijk, (*Report to the Government of Brazil on Forest Inventory,* p. 23), in summing up his forest inventory work for Brazil, suggests that "it is probably more advisable to put an effort into the preparation of management plans for reforestation or conversion pinheiro stands near the main highways and large cities, than to continue with general surveys which, in the end, will only confirm that the natural grown pinheiro forests are overcut."

gram to bring information about legitimate investment opportunities to bona fide investors.[29] Such a service could be (and is in some cases) provided on a national basis for the large countries, in the same way that the large multinationals have means for analyzing their own opportunities. However, there are many smaller and medium-sized countries and investors with interests in the forest-based sector, and some form of international service might be the most logical approach for them (in the same way that this type of approach has been suggested for export development programs).[30] Some international investment advisory services exist at present (for example, the Industry Cooperative Program in the Food and Agricultural Organization (FAO)), but a more specialized and effective sector approach seems warranted. Such an approach could also help to develop the expertise and technical knowledge needed in forest investment negotiations; and it could develop a kind of regional clearinghouse for foreign investment information on the forest-based sector. Aid along these lines has been provided to some countries by the Forestry Department of the FAO and other international groups, but more is needed.

Efforts in the directions suggested above have also been made by Latin American countries themselves. A proposal for the establishment of a regional information center in relation to governmental policies on international investment was recently endorsed by the Steering Committee on Foreign Private Investment in Latin America, a group sponsored by the Organization of American States (OAS) and the Inter-American Development Bank (IDB). With similar purposes, the same committee also established a Study Group on Technology. Since the OAS/IDB Steering Committee brings together representatives of the Latin American public and private sectors, the international business community, and international agencies concerned with Latin American development financing, its impact could be important.

While the lack of communication of technical knowledge of local conditions has been a cause of failures in some foreign natural and plantation forest-based projects in the past, in other cases of failure, information and local technical advice were offered, but ignored by the foreign companies. One reason given for such disregard of local advice is that the company finds it difficult to discriminate between qualified and unqualified local persons and dishonest promoters. In the past

[29] Cf. S. Preston, "Information Requirements for Expanding Markets for Tropical Woods," and references cited in his paper.

[30] For example, through the proposed United Nations Tropical Timber Bureau or regional centers (cf. D. Szabo, "U.S. Statement on Export Development," and United Nations Conference on Trade and Development, Trade and Development Board, *Establishment of Tropical Timber Bureaux*.

legitimate investors have been misled by promoters; sometimes substantial sums of money have been lost, and companies have taken this lesson to heart.

Third, if countries want new activity and investor interest is not forthcoming, it is in their self-interest to take the initiative and to spend resources in generating new information which can be used in attracting new foreign, as well as local, interest in forest-based investment opportunities. While general forest inventories and sector studies are obviously helpful and necessary from a planning point of view, countries need to balance this type of survey work with more specific investment- or project-oriented information for areas that appear attractive.[31] The extent to which the government can pass on the costs of such activities to the investor depends on what the results of the information-generating activities show in terms of potential and the amount of competition they can generate for a given opportunity, that is, what the market will bear. Such government information (and publicity) activity is quite common in many developed countries through investor services provided by power companies, state and local economic development departments or agencies, and various other agencies and associations having an interest in attracting investment into a given area.[32]

Often interest already exists in a given area, and the potential investor approaches the country for an initial look at a given investment possibility. When this occurs, it is common for the foreign company to pay for preinvestment studies, receiving in exchange for the information generated a first option to develop an acceptable project under the general policies governing forestry and foreign investment projects. However, unless government professionals are working along with the company, it is difficult for the HC to determine the accuracy of the

---

[31] Herfindahl (*Natural Resource Information,* p. 151) suggests that government information programs "should provide data and help contribute to the formation of a steady flow of *investment projects* in the industries concerned, or contribute to a reduction in the cost of managing and using certain natural resources" (emphasis added).

Heinsdijk (*Report to the Government of Brazil,* p. 21), who headed up one of the largest and most costly forest inventory projects in Latin America, suggests that, "It has been felt for some time that there was *not much point in continuing general forest inventory work* in Brazil. . . . Plenty of general surveys are available of the most accessible parts of the Amazon Region. With the help of aerial photographs the promising parts of these mostly uniform forests can easily be located. . . . When the need arises, a little additional mapping and inventory work will be sufficient for the projecting of management plans, without going to all the trouble of repeating general or pre-investment surveys" (emphasis added).

[32] We are merely emphasizing an approach which has also been used quite successfully by a number of Latin American countries in the past for other sectors.

resulting information and the extent to which all relevant information has been divulged.

In both cases—where government generates and pays for data or where the potential investor does—the other party generally finds it desirable to carry out its own check on the resulting information. This results in added expense, but is necessary in generating an atmosphere of mutual trust and a common base for negotiations. In two cases studied, inadequate checking by one party or the other contributed to later conflict. In both, the HC and/or FI were reluctant to admit this publicly. Yet mutual trust must be built on the recognition that each party has its own interests in mind and they do not necessarily coincide. Without such recognition (and respect for the other's right and need to check), the chances of conflict are increased substantially.

It is difficult to generalize about the priorities for information generation related to expanding activity in the forest-based sector, both because information requirements differ by type of project, location, requirements, the potential investors, etc.,[33] and because of the costs of information for reducing uncertainty will vary in each case, depending on existing information, accuracy of such information, and on the type and size of project.[34] Priorities obviously depend both on costs and returns from such activities. Flexibility is needed, both depending on changes in conditions for a given potential over time and on the type of project.[35]

Finally, once information is generated on a particular area or problem, the mechanisms and institutions have to be ready to implement the needed changes brought to light by analysis of the improved or added information. As an example, one of the major problems facing Latin American forest-based development is inadequate transport infrastructure and high unit transport costs. Studies dealing with this problem have uncovered or emphasized a number of specific means

[33] There is no need to repeat here the types of detailed information required for various types of forest-based projects. Cf. UN, FAO, *Report of the Headquarters Meeting;* Overseas Development Administration, Economic Planning Staff, *Project Data Handbook:* Sect. 7, "Forestry and Wood-Using Industries"; Organization of American States, General Secretariat, *Physical Resource Investigations for Economic Development;* and G. R. Watt, "The Planning and Evaluation of Forestry Projects."

[34] Nelson (*The Development of Tropical Lands,* p. 186) also suggests that it is difficult to generalize about the cost per unit area for accumulating an acceptable volume and quality of data for an investment decision, "since location, economies of scale, the amount of data already on hand, the quality of personnel, and the new techniques available will have a significant impact on unit costs for a given level of precision."

[35] This is one of the points emphasized by Herfindahl, *Natural Resource Information,* p. 191.

for reducing transport cost incidence in total costs through improved efficiency or productivity.[36] On the side of the companies, production and shipments should be planned to take into account the economies involved with unit loads, lower charter rates available in many parts of Latin America, and packaging improvements. The countries, on the other hand, need to review and revise legislation relating to local shipping regulations and to consider investment in infrastructure requirements to accommodate different or modified forms of shipping. The possibilities for cost reductions are quite substantial in many cases through shifts to more efficient transport systems. At present, due to earlier rationalization of transport methods from Asia, it is cheaper to ship wood products from certain Asian ports to the East Coast of the United States than it is from northern Brazil or parts of Central America. In the case of internal transport in Latin America there are similar inefficiencies which could be reduced.[37]

To effectively take advantage of the various opportunities, the governments need to objectively analyze the costs and benefits involved and undertake necessary infrastructure investments if they appear favorable. In many cases political considerations overshadow the results of objective analysis. While this is understandable, the costs involved need to be recognized in their entirety, including the loss of otherwise viable opportunities for investment.

It would appear that the points discussed above should apply equally to stimulating local as well as foreign investment in the forest-based sector. Further, the question of "who pays" is ultimately determined by government goals and potential investor interest. While it is possible (and perhaps logical in some cases) for a government to push harder for a *foreign* investor to share or undertake the cost of an information program which would benefit both HC and FI, there is little evidence of such discrimination when one compares similar-sized local and foreign companies. Rather, it appears that the larger the investing company the more likely it is that the government will push for private responsibility for information programs. But it is also more likely that the larger the company the more it will insist on carrying out its own evaluation. In such cases, what is critical to avoidance of direct conflict is the willingness of the HC to let the investor go about the task

[36] Cf. State University of New York, College of Environmental Science and Forestry, *Proceedings of a Conference on Transportation on Tropical Wood Products.*
[37] Concerning local transport costs and problems in Brazil, cf. H. Knowles, "Investment and Business Opportunities in Forest Industrial Development of the Brazilian Amazon."

of generating the necessary information on which an investment decision can be made.[38]

We have placed particular emphasis on reducing uncertainty because this is what we found to be an immediate and important bottleneck in the minds of both investors and government officials. Improved technology (changing the production function) is also important, as pointed out earlier, but is more of a long-run proposition. At the present time, reducing the uncertainty surrounding investment possibilities with given physical and technical conditions would appear to have potentials for immediate payoffs in terms of stimulating increased investment activity in the forest-based sector.

Requirements for Productive Negotiation and Administration of Contracts

Once a negotiable situation is created the mechanisms needed to effectively negotiate and administer a contract become important. The basic considerations involved are institutional ones.

Most Latin American countries have a number of official agencies that enter into the negotiation and decision making involved in developing a foreign investment project in the forest-based sector, and this in itself can create a barrier, as was mentioned in Chapter Five. However, even if one accepts the necessity for the involvement of a number of agencies, the red tape and lack of clear policies for any given agency create additional confusion and uncertainty for the potential investor. Much can be done to alleviate this uncertainty by providing the interested parties with clear guidelines regarding the specific interrelationships involved. The lack of simplicity and the confusion is often caused quite simply by the scarcity of trained HC negotiators and administrators who have experience dealing with the skilled specialists retained by FIs. Government agencies react to this void by imposing stringent conditions and building up a confusing net of red tape to guard against their own vulnerability.

A number of Latin American countries need help in negotiating contracts with foreign companies. Aid along these lines has been provided in isolated instances by the FAO. However, countries must be

[38] We mentioned the specific case earlier of a company that wanted to export logs to its home country facilities for testing, but was denied that privilege by the HC. While the country may have quite logical nationalistic reasons for its policy (perhaps based on past experience with foreign investors), it, at the same time, has to recognize the potential loss of an otherwise favorable investment project.

able to hire experienced negotiators directly. And such negotiators should have the proper incentives to develop the most favorable but viable contracts possible for the HCs.[39]

While part of the skill of a good negotiator is developed through practice and through his more-or-less innate ability to react reasonably under pressure and to make tradeoffs in a bargaining situation, a great deal of his skill is also a result of training and sound understanding of the conditions and "rules of the game" with which he is dealing in any particular situation. There have been few attempts to develop these skills in the professionals dealing specifically with forest-based development projects in Latin America. It is only in the past few years that any local expertise in the economics of forestry and forest law has been available for use in negotiations in most countries.

The inadequate output of professional foresters with an economics/law orientation is felt most severely in the Latin American forest services. Lack of FI confidence in local forest services is a cause of uncertainty. As indicated earlier, ineffective administration has been a contributing factor to failure of past projects. Part of the answer to this problem is an increased emphasis on economics/policy and management-related training for Latin American foresters. For at least the next decade or so, much of this type of training has to be obtained from non Latin American institutions until adequate local teaching expertise can be developed.[40]

General Implications for Policy

There is a growing local market for forest products in Latin America and export potentials also appear favorable in the longer run. If private foreign investment is to contribute more to meeting the opportunities and needs and to help in expanding the contribution for the forest-based sector to the region's development, it is essential that the countries of the region and the potential investors reduce uncertainty and conflict by developing the necessary information on which to base investment decisions and by communicating more effectively with each other. If uncertainty is not reduced, we envision an increasing scarcity of foreign activity in the sector, even in countries that want outside investment. Information generated and communicated will

---

[39] On this point, see F. Schmithusen, *Handbook on Forest Utilization Contracts on Public Lands.*

[40] Those few who have this type of training have tended to move into strategic government posts which leave them inadequate time for teaching.

only have a favorable effect on decisions if policies regarding forest exploitation and foreign investment in the sector are clearly defined; for example, those related to land ownership rights and definition of concession agreements.

The countries of the region recognize that there are alternatives to the traditional, wholly or majority-owned foreign, forest-based project, even if technical knowledge, marketing contacts, and perhaps capital to develop efficient forest industries on their own are missing. One possibility which is receiving increasing attention is the joint venture with the minority interest being held by the foreign partner. U.S. firms in the forest products sector have not favored this alternative in the past, but may be forced to do so in the future if they want to get involved in the region.[41] Another possibility is the management contract with a foreign technical group (including foreign companies in the forest products sector) where financing might come partly from bilateral or multilateral loans and partly from local sources. Finally, HC firms or government enterprises could directly hire the necessary technical expertise, recognizing, however, that they will have to pay the type of compensation which will attract capable persons who have adequate experience. Marketing might be contracted to a foreign distributor, or a captive market with a foreign consumer might be established.

The trend in Latin America is toward more government involvement in all natural resource activities, and this includes forest-based projects. For example, a new draft forest law for Peru envisions creation of a government forest industry corporation somewhat along the lines of the already existing national petroleum company (PETROPERU). Honduras is requiring national (private or government) ownership of its basic forest industry by the end of 1974, once organizational problems in the Honduran Corporation for Forest Development have been resolved.[42] As another example, recent foreign investment legislation in Argentina allows only a 20 percent maximum foreign participation in the forest sector.[43]

In such instances, it is obvious that the countries are prohibiting or at least not encouraging direct private foreign investment (nor local private investment in some cases) in the forest-based sector. This is a legitimate political decision. However, a majority of the Latin American countries still want new private investment in the sector, includ-

[41] Current interest shown by North American investors in potentials in Honduras indicate a more ready acceptance of this requirement, provided that the conditions of partnership are clearly stated.

[42] Law 103 of January 1974.

[43] Law 20557, Article 6, November 7, 1973.

ing foreign capital and technical and managerial know-how. The reasons for such policies were discussed in Chapter Three. Even in these cases, however, the degree of control is increasing, and the extent to which foreign capital will have the leeway it has had in many countries in the past is questionable. Quite naturally, the basic policy of these countries is to push through negotiation for maximum advantage without losing a worthwhile opportunity. If direct conflict is to be avoided, governments will need to develop a greater awareness of the investor's objectives, circumstances, and means, and the investor's legitimate need (within a private enterprise system) to make a return commensurate with the risks and uncertainties involved and returns expected from alternative opportunities elsewhere. This includes recognizing the difference between the supply price of capital for new operations with high uncertainty and risk and for those which move into areas that are already well known.[44] As a related point, many countries have policies which prevent foreign investors who acquire existing domestic companies from participating in a variety of benefits given to new investments. However, in some cases—including several studied—the taking over of a failing, inefficient local company can provide greater benefits to the country.

While the countries of Latin America have a number of tasks ahead of them if they want an increased flow of domestic as well as foreign private investment in the forest-based sector, FIs, who obviously also have benefits to gain, need to recognize and adapt to certain fundamental and legitimate changes in the attitudes and conditions in Latin American countries. Among other things, they need to be more flexible with regard to accepting local participation in projects. (Indeed, the implications of the discussion in Chapter Five would indicate that many companies could gain substantially from opening up more to local influence.) They also must recognize the legitimacy of the requirement to process wood locally in the case of export projects and the changing HC views on repatriation of capital. The president of ADELA, a knowledgeable expert on foreign investment in Latin America, suggests that one of the major shortcomings in foreign private enterprise in Latin America is "the reluctance to accept change as a necessity for progress, and to adapt to changes."[45] This is true for the forest-based sector as much as for other sectors.

---

[44] The solution here might partly include revised policies with regard to log exports during initial development of forest regions. Further, to the extent that overall costs are higher than in alternative regions such as Asia, governments must be willing to provide compensating incentives to attract investment.

[45] E. Keller, "Opportunities and Problems of Private Investment," p. 5.

References

Aharoni, Y. *The Foreign Investment Decision Process.* Cambridge, Mass.: Harvard University Press, 1966.

Arnold, P. "Why Development of a Tropical Forests Industry Can Be Difficult," *World Wood* vol. 13, no. 11 (October 1972).

Banco Lar Brasileiro. "Doing Business in and with Brazil," prepared by Paul Griffith Garland, 1971.

Bank Exspor Impor of Indonesia. *Timber in Indonesia* (n.d.).

Behrman, J. N. "Promotion of Private Foreign Investment," in R. F. Mikesell, ed. *U.S. Private and Government Investment Abroad.* Eugene: University of Oregon Books, 1962.

Carnoy, M. *Industrialization in a Latin American Common Market.* Washington, D.C.: The Brookings Institution, 1972.

Chudnoff, M. "Research Needs," in *Proceedings of the Conference on Tropical Hardwoods.* Syracuse: State University of New York, College of Forestry, 1969.

Ciriacy-Wantrup, S. V. *Resource Conservation: Economics and Policies.* 3rd ed. Berkeley: University of California Press, 1968.

Comité de Acción Interamericana de Colombia. *Opinions Regarding Foreign Investment in Colombia.* Bogotá: Acción Interamericana, 1971. (Mimeographed).

Davis, O. and Kamien, M. "Externalities, Information and Alternative Collection Action," in R. Haveman and J. Margolis, eds. *Public Expenditures and Policy Analysis.* Chicago: Markham, 1970.

Dean, J. "Measuring the Productivity of Capital," *Harvard Business Review,* vol. 32, no. 1 (January–February, 1964). Quoted in Mead, 1960.

DeBoer, D. "Impressions of Industrial Problems in Tropical Forestry," Organization for Tropical Studies (1969). (Mimeographed).

dosReis Velloso, P. Memorandum from the Joint Brazilian/Council of the Americas Meetings from U.S. Department of State *Airgram* (Unclassified No. A-151, dated June 13, 1972), from the American Consulate in Rio de Janeiro to the Department of State.

Downie, J. *An Economic Policy for British Honduras. Report to the Government of British Honduras.* London: Her Majesty's Printing Office, 1959.

Dunning, J. H., ed. *International Investment*. Baltimore: Penguin Books, Modern Economics Readings, 1972.

Echeverria, G. L. *El Financiamiento Extranjero y el Desarrollo Economico de Guatemala*. Guatemala City: Universidad de San Carlos de Guatemala, 1971.

Eklund, R. "Use of Capital, Labour and Forest Resources in the World Forest Industries," *UNITAS, Economic Review of Finland*, vol. 45, no. 3 (1973).

Friedmann, W. G. and Beguin, J. P. *Joint International Business Ventures in Developing Countries*. New York: Columbia University Press, 1971.

Gane, M. "Forest Harvesting in Trinidad," *Commonwealth Forestry Review*, vol. 45(3), no. 125 (September 1966).

Government of Chile. "Inventario de las Plantaciones de la Zona Centro Sur de Chile," Instituto Forestal, *Informe Tecnico no. 24, 1966*.

Government of Great Britain. Overseas Development Administration. *Project Data Handbook*. Section 7, Forestry and Wood-Using Industries. London, August 1972.

Government of Indonesia. *Technical Guide to Capital Investment in Forestry*. 2nd ed. August 1971.

Gregersen, H. "The Latin American Contribution to United States Forest Products Imports: Problems and Potentials for the Exporter," *Forest Products Journal*, vol. 21, no. 3. (March 1971).

————. "Export Development Programs for Forestry." Paper presented at the Seventh World Forestry Congress, Buenos Aires, October 1972. (Forthcoming).

Hagenstein, P. "Factors Affecting the Location of Wood-Using Plants in the Northern Appalachians," U.S. Forest Service, *Research Paper NE–16, 1964*.

Hayami, Y. and Ruttan, V. *Agricultural Development: An International Perspective*. Baltimore: Johns Hopkins University Press, 1971.

Heinsdijk, D. *Report to the Government of Brazil on Forest Inventory*. Part I. Rome: FAO, 1966.

Herfindahl, O. C. *Natural Resource Information for Economic Development*. Baltimore: Johns Hopkins University Press, 1969.

Hirshleifer, J. "Where Are We in the Theory of Information?" *American Economic Review*, vol. 63, no. 2 (May 1973).

Hughes, J. "Forestry in Itasca County's Economy: An Input-Output Analysis." Agricultural Experiment Station, University of Minnesota, Miscellaneous Report 95, 1970.

Keller, E. "Opportunities and Problems of Private Investment," *Monthly Bulletin*, ADELA Investment Company, S.A., Lima, Peru. (April 1972).

Knowles, H. "Investment and Business Opportunities in Forest Industrial Development of the Brazilian Amazon." FO/MISC/69. Rome: FAO, 1969.

Lamb, B. "Tropical American Forest Resources," in *Proceedings of the*

Conference on Tropical Hardwoods. Syracuse: State University of New York, 1969.

Lambiase, Th. J. *U.S. Direct Investment Abroad in Paper and Allied Products.* Washington, D.C.: Forest Products, Packaging, Printing and Publishing Division, Bureau of Domestic Commerce, U.S. Department of Commerce, 1972.

Mathews, J. B. "How to Administer Capital Spending," *Harvard Business Review,* vol. 37, no. 2 (March–April 1954). Quoted in Mead, 1960.

May, H. K. *The Effects of United States and Other Foreign Investment in Latin America.* New York: The Council for Latin America, 1970.

McGregor, J. J. "Forestry Concessions in the British Commonwealth Countries," *Commonwealth Forestry Review,* vol. 51, no. 147 (March 1972).

Mead, W. "Long-term Investment Planning for Forestry Development." Paper presented at the Fourteenth Yale Industrial Forestry Seminar, New Haven, April 1960.

Mikesell, R. F., ed. *Foreign Investment in the Petroleum and Mineral Industries: Case Studies of Investor-Host Country Relations.* Baltimore: Johns Hopkins University Press for Resources for the Future, 1971.

Myint, H. *Southeast Asia's Economy.* Baltimore: Penguin Books, 1972.

Nelson, M. *The Development of Tropical Lands: Policy Issues in Latin America.* Baltimore: Johns Hopkins University Press for Resources for the Future, 1973.

Nelson, S. "The Looming Shortage of Primary Processing Capacity," *Challenge,* vol. 16, no. 6 (January–February 1974).

Organization of American States, General Secretariat. *Physical Resource Investigations for Economic Development.* Washington, D.C.: OAS, 1969.

Payne, B. H. and Nordwall, D. S. *A Review of Certain Aspects of the Forestry Program and Organization of Indonesia.* FEDS Field Report 10. Washington, D.C.: U.S. Department of Agriculture and the U.S. Agency for International Development, 1971.

Penrose, E. "Foreign Investment and Growth of the Firm," *Economic Journal,* vol. 66, no. 262. (June 1956).

Piper, J. "How U.S. Firms Evaluate Foreign Investment Opportunities," *MSU Business Topics,* vol. 19, no. 3 (Summer 1971).

Preston, S. "Information Requirements for Expanding Markets for Tropical Woods." Paper presented at the Seventh World Forestry Congress, Buenos Aires, October 1972.

Pringle, S. "World Supply and Demand of Hardwoods," in *Proceedings of the Conference on Tropical Hardwoods.* Syracuse: State University of New York, 1969.

Richardson. "Theoretical Considerations in the Analysis of Foreign Direct Investment," *Western Economic Journal,* vol. 9, no. 1 (March 1971).

Rosenthal, G. *The Role of Foreign Private Investment in the Development of the Central American Common Market.* Chicago: University of Chicago Press, (In press).

Schmithusen, F. *Handbook on Forest Utilization Contracts on Public Lands.* (FO:UNDP/Misc/71/1). Rome: FAO, 1971.

Schuster, E. G. and Pendleton, T. H. "Decisions on Locating a Veneer Plant," *Southern Lumberman,* vol. 217, no. 2704 (December 15, 1968). Quoted in Wolf, 1970.

Somberg, S. *Timber Sales Contracts for Latin America.* Bulletin 23. Nacogdoches, Texas: School of Forestry, Stephen F. Austin State University, 1972.

State University of New York, College of Environmental Science and Forestry, *Proceedings of the Conference on Transportation of Tropical Wood Products, November 16–18, 1971.* Syracuse: State University of New York, 1972.

Stearns, J. L. "The Problems Encountered in Introducing a New or Unknown Wood on the American Market," in *Proceedings of the Conference on Tropical Hardwoods,* Syracuse: State University of New York, 1969.

Stevens, "Fixed Investment Expenditures of Foreign Manufacturing Affiliates of U.S. Firms: Theoretical Models and Empirical Evidence," *Yale Economic Essays,* vol. 9, no. 1 (Spring 1969).

Szabo, D. "United States Statement on Export Development," delivered at the Fourth Meeting of the Ad Hoc Group on Trade to deal with Tariff and Nontariff Barriers and Related Matters, Washington, D.C. October 19–31, 1970. (CIES/CECON/COMERICO/34) Washington, D.C.: Organization of American States, Inter-American Economic and Social Council, 1970.

Takeuchi, K. "Tropical Hardwood Trade in the Asia-Pacific Region," *World Bank Staff Occasional Papers,* no. 17, (1974).

Tuolumne Corporation. *Market Study for Forest Products from East Asia and the Pacific Region;* prepared for the Food and Agriculture Organization of the United Nations. Rome: FAO, 1971.

United Nations, Conference on Trade and Development. Trade and Development Board. *Establishment of Tropical Timber Bureaux.* Document TD/B/C.2/AC.2/19. Geneva: United Nations, August 20, 1968.

United Nations, Department of Economic and Social Affairs. *Foreign Investment in Developing Countries.* New York: United Nations, 1968.

United Nations ECLA/FAO/UNIDO. Papers from the Regional Consultation on the Development of the Forest and Pulp and Paper Industries in Latin America, Mexico, D.F., May 19–26, 1970.

United Nations, Economic and Social Council, Economic Commission for Latin America. *Account of Proceedings and Recommendations of the Regional Consultation on the Development of the Forest and Pulp and Paper Industries in Latin America.* Document E/CN.12/858. Mexico, D.F.: United Nations, June 25, 1970.

United Nations, Food and Agriculture Organization. *Latin American Timber Trends and Prospects.* Rome: FAO, 1963.

———. *World Forest Inventory.* Rome: FAO, May 1966.

———, Pulp and Paper Advisory Committee. *Some Obstacles to a Private*

Investor in Establishing Export-Oriented Forest Industries in the Develop-ing Countries. FO: PAP/66/12. Rome: FAO, October 24, 1966.

———, Committee on Wood-Based Panels. *Obstacles Impeding the Flow of Investment Capital to Forest Industries in the Developing Countries.* FO:WPP/66/11. Rome: FAO, November 1966.

———, FAO/IBRD Cooperative Programme. *Outlines for Projects to be Presented for Financing,* (Provisional Draft). Rome: FAO, September 1967.

———. *Report of the Headquarters Meeting of Forest Inventory Exports on UNDP/SF Projects.* Document FAO/FO:SF/67, IM17. Rome: FAO, October 9, 1967.

———. *1969–70 Yearbook of Forest Products.* Rome: FAO, 1971.

United States, Department of Agriculture, Forest Service. *Outlook for Timber in the United States.* Washington, D.C.: USDA, 1973.

United States Department of Commerce. "Paper Industry Leads in Foreign Growth," *Pulp, Paper and Board Quarterly Industry Report,* vol. 27, no. 3 (October 1972).

———. *U.S. Direct Investment Abroad—1966. Part I: Balance of Pay-ments Data.* Washington, D.C.: USDC December 1970.

———. *U.S. Direct Investment Abroad—1966. Part II: Investment Posi-tion, Financial and Operating Data. Group 2. Preliminary Report on Foreign Affiliates of U.S. Manufacturing Industries.* Washington, D.C.: USDC, 1972.

Vernon, R. "Foreign Enterprises and Developing Nations in the Raw Ma-terials Industries," *AEA Papers and Proceedings,* vol. LX, no. 2, (May 1970).

———. *Sovereignty at Bay: The Multinational Spread of U.S. Enterprise.* New York: Basic Books, Inc., 1971.

Watt, G. R. "The Planning and Evaluation of Forestry Projects," *Common-wealth Forestry Institute, Paper No. 45.* Oxford: University of Oxford, 1973.

Wells, D. A. "Economic Analysis of Attitudes of Most Countries Toward Direct Private Investment," in R. Mikesell, ed. *U.S. Private and Govern-ment Investment Abroad.* Eugene: University of Oregon Books, 1962.

Westoby, J. "The Role of Forest Industries in the Attack on Economic Underdevelopment." Reprinted from *Unasylva,* vol. 16(4), no. 67 (1963).

Wolf, Ch. H. *Wood Industry Location Decisions.* Research Series 15. Mor-gantown: Office of Research and Development, Appalachian Center, West Virginia University, May 1970.

Wollman, N. *Water Resources of Chile.* Baltimore: Johns Hopkins Uni-versity Press, 1968.

Yale University, School of Forestry. "Financial Management of Large Forest Ownerships," in *Thirteenth Industrial Forestry Seminar.* New Haven: Yale University Press, 1960.

Zivnuska, J. *U.S. Timber Resources in a World Economy.* Washington, D.C.: Resources for the Future, Inc., 1967.

Appendix I **U.S. Direct Private Foreign Involvement in The Forest-Based Sector of Latin America**

The following list shows U.S. corporations that were involved in the forest-based sector of Latin America beginning in 1971. Some of the corporations were only involved in export or conversion activities and not in the direct processing of wood. Further, some of the paper companies were utilizing non-wood-fiber pulp and/or wood fiber pulp imported mainly from the United States. The list is based on country information, personal interviews, company reports, U.S. Department of Commerce Trade lists, United Nations information, etc. It is believed that it includes nearly all major U.S. involvements in the forest-based sector. In addition to the firms included, we identified some twenty-two U.S. individuals, or groups of individuals, who also had investments in the sector in Latin America. Finally, at the time this list was prepared, information was not available for Guyana, Surinam, British Honduras, and Panama.

After details had been collected on the sizes, products, raw material sources, etc., for each of the listed corporations, a sample of companies was selected for in-depth interviews. (See Appendix II and Chapters One and Three).

U.S. Corporation	*Countries*
American Can Company	Brazil
Astoria Panamericana de N.Y.	Peru
Balsa Ecuador Lumber Corporation	Ecuador
Bear Murphy Incorporated	Brazil
Bemis Company, Inc.	El Salvador
Bethlehem Steel	Brazil
Boise Cascade Corporation	Colombia, Guatemala
Cadmus Intl. S.A.	Nicaragua
Cannon Craft Company	Mexico
Champion International	Brazil, Honduras, Peru
Celanese Corp. of America	Mexico
Container Corporation	Colombia, Mexico, Venezuela
Continental Can Company	Brazil, Colombia

Continental Shellmar International Co.	Venezuela
Crown-Zellerbach International Co.	Chile, Costa Rica, El Salvador
Dixie Wax Paper Co.	Mexico
Evans Products Co., Int.	Nicaragua
Georgia Pacific Corporation	Brazil, Ecuador
Gillespie and Company, Inc.	Venezuela
Hawthorne Lumber Company	Honduras
Holiday Inns of America	Mexico
Industrias Commercial Minerias S.A.	Brazil
Internal Balsa Corporation	Ecuador
International Paper Corporation	Colombia, Ecuador, Honduras, Venezuela, Peru
Kimberly-Clark Corporation	Colombia, El Salvador, Mexico
Koppers Company, Inc.	Guatemala
Kruger Pulp & Paper	Colombia, Venezuela
National Bulk Carrier	Brazil
National Paper and Type Company	Mexico
North American Rockwell	Colombia
Parsons & Wittemore	Argentina, Brazil, Chile
Pascagoula Veneer Co.	Colombia
Potlatch Forest, Inc.	Colombia
Olin Mathieson Chemical Corp.	Brazil
Resources International	Colombia
Robinson Lumber Company	Honduras, Guatemala
Schlumberger Ltd.	Venezuela
Scott Paper Company	Argentina, Colombia, Costa Rica, Mexico
Simpson Timber Company	Chile
Sonoco Products Company	Mexico
Southwest Moulding Company	Mexico
St. Regis Paper Company	Argentina, Ecuador, Nicaragua
Standard Fruit and Steamship Co.	Costa Rica, Honduras
Timberland Lumber Company	Honduras
Vincent and Wellch Inc.	Mexico
Westvaco	Brazil
Weyerhaeuser Company	Guatemala, Venezuela
W. R. Grace & Company	Ecuador, Peru

Appendix II **Interview Schedules for University of Minnesota Project Study of Direct Private Foreign Investment in Forestry and Forest Industries**

Companies

Through a grant from Resources for the Future, Inc., the University of Minnesota is undertaking an intensive study of the relationships between foreign investors in the forest industry sector and Latin American host countries. The main objective of the study is to identify and analyze the sources of misunderstanding or conflict between the two, focusing on conflicts and problems related to the renewable nature of forest resources. The goal is to provide some insights into possible solutions to real and apparent problems. Such insights hopefully will be helpful to both host countries and investors.

Four main sources of information will be utilized: (1) interviews with foreign investors; (2) interviews in host countries with management of affiliated companies; (3) interviews with host country government and private officials; and (4) observation in, and documentation available from, the host countries.

The attached outline is intended for use in the interviews with corporate officers of the investing companies. It is being sent to you at this time to help identify which company officer or officers should most appropriately be interviewed and to minimize the time spent in the interview process itself.

In keeping with well-established practice, anonymity is assured for information provided by individual companies.

Interview Outline for Inquiry Among United States Companies with Operations in Latin America

I. *Basic Facts on Company's Latin American Operations*
   A. In what countries is company involved in Latin America?

109

For Each Country

B. When and in what form was the local operation organized?

C. What is the nature of the product produced?

D. Reason for establishment (i.e., entry into local market, support U.S. operations, entry into third countries, etc.)

II. *Control and Ownership of Operations*

A. Type of affiliation (subsidiary, joint venture, etc.)

B. Extent of control (percentage of ownership)

C. Approximate amount of capital invested in operation (we only need rough order of magnitude) and most recent gross sales receipts.

D. Extent of autonomy of local manager in dealings with government [what types of (policy) decisions are left to the local manager or other chief administrative officer, as the case may be?].

E. If operation is integrated back through primary processing of wood, to what extent is raw material source company owned, and to what extent is reliance placed on local government and private sources?

F. If your operation is a joint venture with a local company, what do you perceive to be the advantages and disadvantages? If not a joint venture, has this in any way worked to your disadvantage?

III. *Special Treatment (Positive and Negative) from Host Country*

A. If operation involves primary processing of wood raw materials, are there any special host country regulations, laws, or incentives which affect you as a foreign investor?

B. If operation is in one of the Andean Pact countries, how do you perceive the influence on the operation of Decision No. 24, "Common Treatment on Foreign Capital, Trademarks, Patents, Licensing Agreements and Royalties," approved by the Commission of the Cartagena Agreement in 1970?

C. Is operation discriminated against in any way because it is a foreign-owned or -controlled project?

D. Any problems with transfer of earnings or repatriation of capital?

E. Has operation been able to take advantage of special tax incentives, tariff exemptions, etc.?

IV. *Host Country Conditions*

A. What is the nature of the taxation issues faced?

B. What is the nature of the labor issues faced?

C. Export problems (e.g., pricing problems for exports to parent company).

D. What is the nature of the government attitude toward the operation and, in general, toward the category of business in which the operation would be classified (e.g., pulp and paper, lumber, plywood, etc.)?

E. Have you in the past faced, or do you now face, shortages of infrastructure (e.g., public utilities, public services, etc.)?

V. *Investment Decisions*
A. What were the two or three most important factors which led you
to undertake operations in the Latin American countries in which
you are operating?
B. Did you consider alternatives such as purchase from a local company or licensing arrangements with local producers?
C. What were the main factors responsible for any major expansions
you have undertaken in your operations in Latin America?

VI. *General Questions*
A. General appraisal of the local conditions in host country for
foreign investment in forestry and forest products operations.
B. Any additional significant points which have not been covered in
previous questions.
C. What are major contributions of operation to host country develop-
ment?
D. Suggested solutions to major problems faced in Latin America.
E. What are the two major problems you face in your Latin Amer-
ican operations?
F. If timber is being cut in host country, what are the main condi-
tions of your concession agreement (or is land owned by company)?
Is a copy of the concession agreement available for our review?
G. What percentage of employees (professional and skilled) in Latin
American operation are from the United States?

VII. *Would you be willing to give a letter of introduction to your manager
in the Latin American operation when we commence our field research in the host countries?*

Host Countries

The following background on a research project currently under
way at the College of Forestry of the University of Minnesota will
provide the basis for forthcoming interviews to be arranged with the
purpose of obtaining additional information on the subject.

Background on Study

In 1971, Raymond Mikesell and associates published a study of foreign
private investment in the nonrenewable natural resource industries (petro-

Note: This questionnaire has been freely translated from Spanish.

leum and minerals). That study, also financed by Resources for the Future, Inc., attempted to look at the experience of foreign investors in the minerals and petroleum sectors, primarily in terms of the relationships or conflicts between investor and host countries and with emphasis on the resource-related aspects of such investment. The present study is a parallel effort to look at the experience of investors in the forestry and forest products sector, with specific emphasis on problems associated with foreign investment in *renewable* natural resources, and the contribution of such investment to development.

Specifically, we are interested in analyzing the impact of foreign investment in the forest products sector on forest development and forest management in less developed countries, the types of relationships involved, the problems encountered, the ways in which such problems have been overcome (or have not been overcome), the contributions of the foreign companies to improvement in and increased utilization of the forest resource base, and the contributions of such companies to general economic development of the host countries.

Information Requirements

The following outline lists the types of information which are relevant to our study. To the best of our knowledge, none of the questions relate to "confidential" factual information (in the sense that it is not available to the public). In any case, it is important to mention that we are interested in general trends in foreign investment and its contributions, and therefore the information will not be identified with institutions or individuals if they desire not to be identified.

I. *General Background Information on Governmental Policies Toward Foreign Investment*

We are interested in the most important economic policy measures adopted by the government to afford incentives or to restrict the activities of foreign corporations in the forestry sector. This includes, for example, tax incentives, characteristics of concessions, limitations imposed on foreign ownership, price control, etc. We would like to obtain copies of the relevant laws and regulations.

II. *Main Contributions of Foreign Corporations to the Development of the Host Country*

Foreign investment may contribute to the development of the host country in a number of ways: for example, it may contribute to foreign exchange earnings, technological development, access to international markets, employment and training of the local labor force, improvement of forest capital and management, etc. We are interested in identifying the specific contributions to your country, as you see them.

III. *Misunderstandings Between Government and Foreign Corporations*

Here we are interested in the origins of problems between your government and foreign corporations. We are also interested in the measures adopted to solve such problems.

IV. *Modifications in the Government Economic Policy Deemed Necessary to Deal with Foreign Investment*

Library of Congress Cataloging in Publication Data

Gregersen, H   M
   U.S. investment in the forest-based sector in
Latin America.

   References: 5 pp.
   1. Forest products—Latin America. 2. Investments,
American—Latin America. 3. Forests and forestry—
Economic aspects—Latin America. I. Contreras,
Arnoldo, joint author. II. Title.
HD9764.L32G74   338.4'7'674098   74-21754
ISBN 0-8018-1704-8